Everything You NEED to Know About TIG Welding

Shawn J. McDonald 2018

About the author

Hi, thank you for purchasing this book. My name is Shawn McDonald, and I have been welding for about 14 years now. I started off in my parent's garage with a craftsman 60amp wire feed 110v welder that I bought to try and shave the door handles on my 1987 Chevy s10. I started from there and continued working my way up the ranks, and eventually went on to do ASME boiler/pressure vessel welding, and lots of high end fabrications. I wrote this book because I truly enjoy helping people perfect their craft, and I felt like I may be able to contribute my own experiences and history to help someone just learning the trade better figure things out. No welder is ever done learning, so here's my personal take on TIG welding. And if you like my book, please be sure to check out my website and YouTube channel at www.welding-tips.net

Introduction

Of the 3 most common types of welding, (SMAW, GMAW, and GTAW), GTAW, Gas Tungsten Arc Welding also known as TIG welding, (or Heliarc if you're talking to someone old school) is the most versatile of the welding processes. It can be used on any material, it produces the cleanest and best penetrating welds, and it allows you much more control over how the weld lays down. With MIG welding or stick welding, the filler material acts as the electrode, meaning its constantly feeding filler material into the puddle. With TIG, you have the option of slowing down, using less filler, and working the puddle to achieve the look and size weld you need.

In addition to control over how much filler is added, you have control over how much heat is being put into the workpiece as well. This can be advantageous in situations where you are bridging a large gap, and you have to add a lot of filler material. With other processes, the weld moves along slowly and starts to overheat. With TIG, you can back off the pedal without losing the arc or gas coverage and cool the puddle down and continue welding. This makes TIG welding great for doing plug welds, filling holes and doing build ups.

Another fun fact is that TIG welds are typically softer than MIG welds or stick welds. Which means they grind easier, and hammer and form easier, so if you are working with some sheet metal and you need to do some hammering around an area you welded, it's much less likely to crack if you TIG welded it. The weld nugget is more malleable which makes it much easier to manipulate.

If you do a lot of ornamental or detail work like picture frames, art work, railings, awnings, light fixtures, that kind of thing, you will most likely be grinding and blending a lot of welds for aesthetics. TIG is the way to go because it gives you more control over the weld, with the ability to input less heat. Less heat amounts to less material distortion, and a better looking overall product, which, when you're working on ornamental or artistic pieces, is critical.

Many people find TIG welding to be challenging. I think the problem is that there is quite a bit about the process that may not be self-evident. The reality is, TIG welding is very simple and with enough practice and a little bit of time, you too can be laying down stacks of dimes every single time.

In this book, I will talk about my approach to TIG welding. I will show you the different things I look at, and different techniques and tips that may help you improve your technique. I spent years learning these techniques, and never spent any time in a welding school.

All my experience and knowledge is from hands on work in the field and having screwed up once or twice in the process. The good thing, is that those screw ups usually come with lessons, and in time, those lessons translate toe experience. If you are new to TIG welding, you'll want to keep reading. And even if you are an experienced welder, it's always nice to see how other people work to, gain outside perspective. In this trade, the more you know, the more you're worth, so take in as much as you can!

Getting Started

So, before we get into the actual welding process, let's talk a little bit about welding machines. There are two main types of welding machines that you will run across today. There are transformer-based machines, and Inverter based machines. Both of these types of welding machines do the same things, but each has their own way of doing it, and its own advantages and disadvantages.

Transformer based machines are the originals. The older technology that's been tried and true, and still is used in most manufacturing settings that do a lot of welding.

A transformer-based machine is basically just a big power source that takes the input voltage, and steps it down to around 16 volts give or take. Transformers are big heavy and expensive pieces of equipment. They are made of coils of copper wire with a shellac coating wrapped around a ferrite core. The coil is then tapped at various locations to give you different output voltages.

Transformers are used in all kinds of equipment even to this day, However, there is an increasing number of welding machines and equipment in general being constructed with what's known as Inverter technology.

Inverters are circuits that take standard AC line voltage and step it down to an appropriate DC voltage for welding. You are probably familiar with an inverter that you would put in your car to turn 12v into AC power, so you can run tools and appliances in your car?

This Is very similar, except it works the opposite way. In an inverter you would use in your car, it's taking your battery voltage, which is a low 12-14v DC, but has a large supply of current, And steps it up to 110v AC power. So, you get a step up in voltage, But the trade off is you have much less available current. And typically, there will be some type of oscillator circuit inside of the inverter to take that DC power, and oscillate it into AC. This is mostly all done with transistors, and other modern components.

An inverter-based welding machine works very similar except you are starting with the larger AC voltage as your input and using a series of transistors as voltage regulators to step down the voltage. And you get the opposite effect as you would with a 12v-120v inverter – more current. This is due to ohms law, and how voltage, current and resistance correlate to one another. If you have 220v at 50 amps as a supply voltage, and you step it down to 12v, you have all this extra potential that becomes available as current. So, your machine can effectively put out 200 amps of current with a 50-amp input.

This is the same basic thing that a transformer does, except it does it using solid state electronics. Transistors.

The benefit of using a transistor-based circuit is that it's low cost, Transformers are typically made of large amounts of copper and iron, which adds up in price, and ultimately makes the machines much more expensive. Second, is consistency. A transistor-based circuit can be much more accurate to what you set the machine to.

On a transformer-based machine, the output of the transformer is always a ratio to the input. You have a primary coil and a secondary coil. Depending on whether it's a step down or a step-up transformer, the primary and secondary are wound differently.

So, for example on a step-down transformer, you will have a primary coil that is wound to a certain specification, using a thinner wire and more wraps. And since you will be stepping down the voltage, and thus increasing current, the secondary will be wrapped with less wraps of wire, with a thicker diameter.

There are taps throughout the secondary winding. For different applications. These reflect different amounts of windings, so for example say your secondary coil has 30 turns, there may be a tap at 15 turns, a tap at 30 turns a tap at 25 turns. Each of these will have a different output, But the output will always be a ratio to the primary coil.

So, what does all this mean? Basically, your wall voltage will affect your welding voltage. If you put a true 220 volts into the transformer, you will have an output that is whatever ratio the transformer is tapped at.

If you have a supply voltage that is not so stable and let's say some days it puts out a true 220, But maybe when you run the compressor, it drops to 210v. Well that 10-volt drop will translate across the transformer, and the output current of the transformer will drop accordingly.

Since a transformer machine is basically just a big analog power source, that current drop translates directly to your welding current.

This is not so much the case with an inverter-based machine. Inverters are made with silicon transformers, which are arranged into smaller circuits such as voltage dividers, and voltage regulators. The significance of this, is that you are more effectively able to dial in the desired parameters.

On a transformer machine, Especially the older ones, your controls are basically just adjusting the output of what comes off the transformer. Really old welding machines, like those old Hobart's that were about the size of a garden tractor would have a big rheostat on the front panel to limit the current. A Rheostat is basically a big variable resistor that is capable of withstanding large amounts of power. So that rheostat is limiting the actual welding current before it gets sent to the electrode.

And since your voltage and current coming off the secondary windings of the power transformer are based on the supply voltage and current on the primary side, there is really no way to get an accurate reading by just setting the machine to what the number s on the front say.

With inverter-based machines, now we can set all the critical parameters first, and lock those in, so that we are adjusting the supply current using voltage divider circuits and voltage regulator circuits. We know that for a TIG welder to run correctly, it needs to be welding at around 16 volts. If it drops below that, the arc may extinguish, or become very unstable and give poor results.

So, by using solid state technology, we can set the supply voltage and lock it in place, and then adjust the current accordingly. So, the machine will take the supply voltage coming from the wall and regulate it to a steady 16 volts. And so, any variance of the supply voltage to the machine from your wall, would only result in a potential drop in maximum welding current. You wouldn't see a voltage drop like you would with a transformer.

So, what does all this mean? Why does it matter? Well if all you are doing is welding, it probably doesn't to you. But, if you have used both machines and noticed a difference in performance, therefore.

Some people notice that transformer machines tend to have a bit more power output for the same sized machine than an inverter. And this not necessarily because the machine is better, it's that the inverter machine is better at regulating what's coming into it, and better at regulating what it's putting out. You get out what you put in with those older machines.

So, the big advantage that most people will appreciate is the convenience of using an inverter machine. Because we need a lot of current to weld, a transformer-based machine needs a big heavy transformer with lots of thick copper windings to withstand all that current.

Inverter based machines don't. they regulate all the input with transistors, which are cheap, small, and light. The result is more portability, not only in size, but in power input as well. Many inverter machines can run on both 110v and 220v, with a good quality welding arc on both input voltages (for the most part that is, sometimes 110v is crappy in residential buildings) this opens worlds that simply wouldn't be possible with big heavy transformer machines.

It would have been very difficult in the 1960s to have someone come and TIG weld a stainless-steel range hood in a restaurant. They would have had to lug out this gigantic welding machine, and run long leads into the work area, probably use a scratch start setup and a power block tied off to a stick welder.

Nowadays, you can buy a small portable TIG welder that runs on both 110v and 220v, its light, portable and best of all... inexpensive. Companies are now selling these types of machines for around a grand or less. That's pretty awesome when you think about the potential

The downside is that transistors are sensitive. Power surges can be catastrophic and overheating them can also be very bad. But for the most part, they seem to be pretty reliable, even on many of the lower priced options.

Transformer machines still have their place, however. If you are looking for a welder to put in a shop, that's going to see 10-12 hours of welding every day, and you're running It close to the high end of the output, that machine is going to get pretty hot. A transformer machine might handle that abuse a little better than an inverter machine. So, I would look into one of those.

But if you're just a hobbyist, or a small business owner, like myself, who needs portability and convenience, There's no other option than inverter.

I have a setup that's extremely portable that I use for all types of jobs, and it's adaptable, so I can run it in someone's house if need be, or in an industrial environment, doesn't matter. I can carry it, and most of my equipment into the job with only a couple of trips, which saves me time. And best of all it was cheap! And if it breaks, I can totally replace it for cheap too.

Some of the bigger transformer machines like the Miller syncrowave cost upwards of $10-$12,000. If you're a small-time guy, it's going to take a bit of work to be able to cover that cost. The machine I use, when it was new retailed for about $2,000. I can make that back on one job without really breaking a sweat, so it's more than paid for itself dozens and dozens of times over.

But aside from the power input, and the technicalities of how the machines work, all TIG welders have a pretty similar layout. You have a TIG torch, a ground clamp, and a foot pedal or finger switch. Torches vary in size, but most are similar in design. Some will have a gas valve built in, some will have a switch-built in. some have a flex head, there are elongated torches, close access torches, pencil torches, just about any torch configuration you can imagine is out there, however the standardized torches are more than adequate for most jobs.

Torches have a standardized sizing, so all manufacturers of a number 20 sized torch will use number 20 compatible hardware. This is great because you don't have to deal with proprietary consumables that are hard to find. This is an issue with MIG welders but has been improving over time. When Chinese made welders started becoming available on the internet, you'd see them up for sale when the company went out of business and it used some proprietary hardware that you could only order from the same manufacturer, so basically once they went out of business, your welder became completely worthless, unless you found a way to adapt a new gun, and once you spent the money on a new gun and the parts needed to adapt it, you might as well have just bought a new machine.. Luckily Tig welders are a bit more standardized and don't really have those issues these days.

So how does it all work? Well, the foot pedal controls the amperage on the machine, and can be substituted for a thumb wheel or slider on the torch handle, or just an on/off switch as I prefer myself. The pedal is great for sit down work, where you're at a bench, but if you must work in an awkward position, the torch mounted control is much more effective and useful.

There are several modes of operation for these torch mounted remotes, so I always recommend looking into the different types of controls available. I have used some interesting ones. You have thumb wheels, sliders, paddles, pushbuttons, and one that I used most recently was a paddle, but it had 4 clicks which gave you incremental amperage. Somewhat like a torch mounted foot pedal. It did take a little while to get used to, but once I was used to it, I really liked it. The first setting was 25% of your current setting, the second was 50%, third was 75%, and full blast gave you 100% of the amperage you have the machine set to.

I found this useful in fabrication, if I was tacking pieces together, I could give it a full blast and tack my parts pretty fast, and then if I needed to run a bead I could use a little less amperage if needed without having to change the machine. And I could still easily work in awkward positions. But there are other types of hand controllers that are just as functional.

There are simple momentary pushbuttons, which is what I currently have, and there are some that stay on when you press and release the button, and only turn off when you press the button again.

You can get creative with a remote switch in terms of your technique as well. You still have options for tapering off the arc on many machines. You can play with downslope and upslope settings on the machine to do manual pulse welding where you get full current if you hold the switch, but when you let go the amperage starts to taper off, then you pulse it again to keep the puddle flowing. I've found that useful in situations, especially where you are filling a hole, or a gap and you don’t want the puddle to get too hot and drop out on you.

There are the thumb wheel style controllers that are popular. They will have a wheel that you roll your thumb across to ramp the voltage up, and on the down side, there is a click in the wheel to shut the arc off.

Sometimes they are a bit clumsy to use in my opinion. Sometimes you must slide your thumb across it a few times to get it up to full power, which obviously takes a little longer, and gives you the potential to move the torch the wrong way and maybe dip your tungsten if you aren't careful.

The way I liked using the thumb wheel was having the machine set so that as soon as you clicked the wheel out of the off position, the amperage was at 100%. So basically, it worked as a click on, click off type of controller.

However, Miller sort of had a good idea with their cheapest machine, the diversion. The torch handle is cool, the rest of it sucks in my opinion it has a short lead with a rubber covering, and non-removable from the machine, YUCK. I much prefer the option to change out my hoses for longer ones if I need to, and to have the ability to remove the torch quickly using DINS connectors, for when I need to carry my machine into a job. It makes it much easier to carry, and less likely to damage the torch in the process.

The diversion has a nice torch handle though that has a built-in amperage wheel, and an on/off button. I would love to have a torch handle like this on my personal machine, if it were available in a number 20, or a number 9 water cooled size. The reason I like this, is because you can use the wheel to preset your ramp in current, and then turn the arc on, and its right there. no fumbling with the roller to get it up to temperature. And when you're done, you click it off.

It works great for fabrication. If I'm tacking, I will roll the wheel up to the highest setting, and crank the machine, and just click on and off pretty quickly to give a quick pulse of weld current, Then when I want to weld, I can back the wheel off a bit to where it's more useful, and still have the ability to come up on the power a little if need be.

I'm surprised I don't see a design like this in more industrial TIG welders. It's extremely convenient once you get used to it. I would recommend you investigate one of these torch mounted finger controls right from the get go, because in my opinion, they teach you how to weld with proper technique, since you don't necessarily can always vary your current on demand.

Beyond the basic layout of the machine, the first step to producing high quality welds is making sure your machine is setup correctly for the material you are running. Most materials will be welded using DC current. However, some like aluminum and magnesium require AC current to be welded. So, your first step is going to be to determine what material you will be working with. Your TIG machine may only be DC capable, which will limit you to doing steels and stainless steels for the most part.

Most machines are similar, and you will want to get the manual for yours to figure out specific settings, but they all have a pretty similar basic layout. There's going to be a process selector (dc+, dc-, or AC on some machines DC process is selected by reversing how the leads are connected to the machine. Some have just a switch you flip.), an amperage control, usually some type of mode selector for stick or TIG, and usually a switch for high frequency or lift arc.

Many inverter-based welders have several options to allow you to better dial in your arc. These will include pre/post flow gas, which allows you to adjust the length of time the shielding gas is on before and after the arc starts and stops. Post flow is usually more commonly found. Even on old transformer machines, but not all machines have pre-flow. Post-flow is helpful in reducing pinholes and crater cracks on arc stops.

If you don't have enough post flow gas going to the weld when you let off the arc, what happens is the weld doesn't cool fast enough, and it will start to absorb contaminants from the surrounding atmosphere. It usually won't give you full blown porosity, but it can leave a pinhole in the middle of your weld. These pinholes are prone to cracking, and no QC inspector or weld inspector worth a damn will let it pass.

So always be mindful of your post flow. Of course, you can have too much post flow, where you are just wasting gas. The amount of post flow you use is going to depend on how many amps you are welding at, and what material you are welding, as they all dissipate heat at different rates.

In addition to pre-and post-flow gas controls, many newer welders will include a pulse feature. What this does is pulse the arc at a specific frequency to keep the heat input down. On some older machines the pulse feature allows a very limited number of pulses per second, which makes it more useful in dabbing neatly stacked welds, however with many newer inverter machines, you can pulse up to 150, 250+ pulses per second. The advantage to this, is it allows you to maintain penetration while keeping heat input down. This is useful on stainless steel where warping is a big concern.

Some more full featured welders will have different settings for wave balance on the ac setting, which allows you to adjust the arc to get more cleaning action vs, more penetration.

The difference is in how the arc focuses on the material. When you have more cleaning action set, the arc is going to be 'wider' and won't penetrate as deep down into the metal. the wider arc causes the oxidation layer on aluminum to break up a bit. Now what's really going on, is not just the width of the arc, but how the electrons are moving across the material.

To understand, you need to picture a sinusoidal wave on an oscilloscope. This represents the electricity used to make your weld. On the oscilloscope, you have an X and Y axis, and the waveform oscillates from one extreme to another. This represents the reversal of polarity.

If you know a little bit about electricity, AC is Alternating current. Which means that it switches polarity several times per second. So, at 60 cycles(Hz), which is what most AC line voltage runs at, the polarity reverses 60 times per second.

So, on an oscilloscope, if you were looking at a sine wave at 60hz, the wave would start on the X axis, (the horizontal line) and begin to rise as it moves across the X axis. At some point it reaches a peak. And begins to reverse. This is called the amplitude. Anything above the X axis is considered positive amplitude, anything below it is negative amplitude. Left to right represents time. Since we are talking about 60hz, that means 60 cycles per second, so if you had a waveform with one positive peak, and one negative peak, time would be one sixtieth of a second.

So, a perfectly balanced waveform would have its X axis located in the middle of the waveform, so the negative and positive peaks are the same distance away. X represents the neutral point.

When you move the knob for your AC balance, you are basically changing the point at which the waveform intersects your X axis. On the screen it would look like you moved the waveform down or up, respectively. What this means is that the waveform spends time as positive, or negative depending on how you set the knob.

So, on an atomic level, you have electrons jumping between the base material and the electrode. And they switch polarity 60 times per second. By adjusting this balance, the amplitude of the waveform changes, and it will ether spend more time with the electrons jumping from the workpiece to the electrode, or more time with the electrons jumping from the electrode to the work piece, respectively.

So, you are welding at 16 volts in AC, the X axis on the scope represents the 0-volt mark, the peak on the Y axis, (above the X axis) represents +16v, and the peak on the bottom (below the X axis) represents -16v. you have a 32-volt potential between +16v and -16v.

That 32-volt potential stays intact, but the ratio changes, so if you have waveform set more negative, you might have +8 volts, and -24v. or vice versa. It's still a 32-volt potential, but the waveform is either more positive, or more negative.

So, in electricity, electrons always flow from negative to positive. The electrons have a negative potential, and they are always trying to flow towards the positive potential.

So, when you are welding DCEN, or DC electrode negative, the electrons, are jumping off the tungsten and the work piece is positively charged. This creates the arc force needed to penetrate deeply into the steel.

With aluminum you have a heavy oxide layer that doesn't melt easily, nor does it melt into the aluminum puddle. So, when you run the machine in AC, half of the time the electrons are jumping off the part to the tungsten, which causes the oxide layer to break up, and then when polarity switches and the electrons jump from the tungsten to the part, that oxidation is not there, so It can melt the material and penetrate.

So, adjusting the wave balance makes it so that the electrons spend either more time jumping off the base material, or more time jumping off the electrode. You'll have more penetration when they are jumping off the electrode, but you will get more of the cleaning effect when they are jumping off the base material. As a result of running more electrode positive, the arc becomes a bit 'wider'as its jumping off the base material, and not off the focused tip of the sharpened tungsten. I hope I explained that easily enough to be understood. Its sometimes challenging to explain these things without pictures to show how it all works but back to TIG welders…

Some TIG welders will have a selector to allow you to use a foot pedal or a finger switch. I recommend getting a torch mounted switch or amp controller, because in the long run it will make you a more versatile welder. Relying on the pedal gets you too used to working sitting down and makes it difficult to work out of position. It also allows you to be lazy in selecting your heat.

There will be times where you simply cannot use a pedal, and if you are not used to welding at a set currant, it may be difficult for you to get things done. It's difficult sometimes to weld something with a foot pedal while you are standing up. All your weight ends up being on your opposite foot, and your leg can cramp up, its just unnecessarily difficult.

Also, a lot of people just set the machine hot and use the pedal to correct it. This is not the proper way to do it in my opinion, and if you want to make great looking welds, you need to abandon that practice. You need to be able to get the material to the pinpoint temperature where the metal flows best. This is what’s going to give you the best color, the cleanest weld, and the nicest looking beads.

It’s also good to get you familiar with how much heat you are using on a material. It’s critical to know how much heat to use, especially on thinner materials, where you don't get a second chance. If you try to start an arc with 100 amps on a piece of 22ga. it’s just not going to be your best day of the week.

Learning how to use a finger controller gets you used to the exact temperatures that a given thickness of metal will weld best at. So, if you are ever in a situation where you have to weld a piece of 1/8” stainless, in a confined area where you cannot use a pedal to correct your heat, you’ll know to start at around 70-75 amps instead of 120. The welds will come out better, and you will minimize your chances of overheating the material and crystallizing it if you are in too awkward of an area to move at a fast travel speed.

Aside from that, the machine setup is basic. My personal preference for machines lately has been the newer inverter style machines. I learned how to TIG on a Syncrowave 180SD and have owned a couple old transformer machines in the past. It's really a personal preference.

For me, the biggest advantage to inverter machines is portability. Mobile TIG welding wasn't really a very practical thing years ago, you could run a lift arc setup off a welding rig, but you were limited in what you could do. AC welding is out, unless you carried an Hi freq arc starter, and you were limited to as far as your leads could reach.

Now you can get a good TIG welder that runs on 110 or 220v power and does ac/dc with pulse and wave shape, post flow, etc. everything you need. and you can fit it in the back seat of your car and carry your whole setup from the truck into the jobsite in pretty much one trip.

I've made some good money doing stainless work with a portable TIG setup. Around my area there are a lot of craft breweries popping up, and they all need welding to get setup and running. Most of the brewing equipment is made in china, and isn't really engineered to be user-friendly, so I've had to extend some pipes, install some flanges, steam vents, moving things around mounting brackets etc.

But you can do so much with this setup. Mobile wheel repair, boiler repair, on site fabrication, railings, restaurant and kitchen equipment repairs, you name it.

Presently the machine I use most is an HTP InverTIG, And it's a great machine. I love it. It runs on 110v and 220v, it's got post-flow, pulse, etc. and has a nice CK Superflex torch. So, I can easily replace parts and consumables, they're stocked just about everywhere. its durable and light weight, I can bring it anywhere. The whole chassis is metal, unlike a lot of newer machines. Mine is a bit older, from when inverter machines were new. But its still a great welder, and I have no complaints.

For my mobile welding business, I carry an 80cf argon tank and a few different extension cords because you must be able to adapt to whatever jobsite you're working on. But it's great because if I can't get 220v where I need to weld, I can still get a 110v hookup. And it still has enough nut to weld almost anything I throw at it. A couple of times, I've plugged in at places that had crappy supply voltage, and it gave me issues with the high freq. But I think that was only at one, maybe two places I've been to. So, it's not a huge deal. If it's really that bad where you can't run the machine, and you don't have a

220v outlct available anywhere nearby, you always have the option of running a generator, and since these things take such a small amount of electricity to run, you could use a relatively small generator and still have good results

You just don't have that option with a transformer machine. They're too big, and they use way too much power, you pretty much need a forklift to move one, and most of the larger industrial sized units require at least a 50-amp source.

There are some smaller ones that use 30-amp input power, which is ok for the home user, but in an industrial environment, you probably won't find many welders that run of 30 amps. So, if for some reason you had to bring one out of the shop and weld something in the field. you would need a pretty serious generator to run it if you couldn't find the proper power hookup.

But I do like transformer machines for doing heavy work, and for doing production work where I have to weld 100 or more of something in a large batch. If I have to weld up 100 parts that require me to weld at 100 amps, and I'm using a syncrowave 350… I can weld all damn day until the cows come home, and that machine won't even get warm.

Now on my invertig, at 100 amps, all day long, it might come close to duty cycle if I'm cranking on it long enough. So, the transformer-based machines still definitely have their place.

So back to machine setup; As far as your hardware and consumables go, you have a few options. In my opinion, the best all-around TIG torch is a number 20.

The issue most will find is that this is a water-cooled torch, and you need to have a water cooler to run it. If your machine does not have a water cooler built in, you will need to buy a separate water cooler and connect it, which makes the machine a bit less portable.

There is an air-cooled version of the number 20 called a number 9, but it's only rated at 125 amps, so you're limited in what you can do with it. They are however, the best sized torches for doing finely detailed TIG welding. You have excellent hand control, the consumables are small and light, and easy to work with, and it's a commonly used size. I personally use a number 9 with a flex head for a lot of my mobile work, although if I must go over 100 amps, I do have a number 17 torch head on my truck at all times, that I will switch out, so I don't burn up my number 9.

The most common size found on most store-bought machines is a number 17. This is a good torch as well, it's just a little bigger and can be hard to get into tight spaces. You have a few options for consumables, but the standard gas lenses are big and bulky for doing detail work. They are probably great for walking the cup on pipe, but they can be bulky and hard to get into Tight spaces when you're dealing with fabrication. Eventually you will end up being able to work with it, But I truly feel like it's a limiting factor in being able to make beautiful welds out of position. I know I've been in several situations where I have had a number 17 torch and only a standard gas lens, and it's been a hinderance because I know I could do such a better job if I could reach a little better with a smaller cup and torch head.

They do make low profile cups and gas lenses for a number 17 torch, but I've personally had trouble finding them locally, which if you're in a pinch and need a certain cup or whatever to reach a joint, it's kind of a bummer.

My recommendation if you are setting up this welder in a shop where it won't be moving, is to invest in a water cooler, and get a number 20 torch. You will be able to run at 250 amps almost continuously, as the water cooler is going to keep the torch nice and cool. The issue with running a number 9 is that if you run it over 125 amps, it's going to get hot very fast, you're going to burn up collets and eventually burn up the torch. Did I mention it's going to get hot fast? You'll feel it through your glove, and it really doesn't take long. Even at 100 amps it still heats up pretty good. So, if you can swing a water cooler, I'd go for it. It's worth the scratch.

If you can't get the water cooler assuming your TIG welder is rated for more than what a number 9 torch is made to withstand, I would suggest sticking to the number 17 torch instead of risking burning up the number 9 torch. If you absolutely need to use a number 9 for the tight access cups, then you got to do what you got to do. There are stubby gas lens kits available, as well as Pyrex cups for the number 17 torch to make it a bit easier to handle, as far as bulkiness. The cups are closer in size to what you'd run on a number 20. So, you can still get a reasonably low-profile torch with a number 17, just you're going to have to order everything most likely.

If you are wondering what a gas lens is, bottom line, you'll want to have them. It's a replacement for the standard collet body and ceramic cup setup that consists of a gas diffuser with a screen, and a wider diameter ceramic cup to help direct the gas in a less turbulent manner towards the weld puddle.

Gas flows through the screen and diffuses it for better shielding coverage. As an added bonus, because they are so much more efficient at delivering gas, you end up using less gas. And because the coverage is so consistent, you can run your tungsten stuck out much further. This means you can get into much tighter spaces without worrying about putting down a gray porous weld.

So, if you don't have a gas lens, I would recommend buying a few. They are specific to tungsten size, so if you switch between tungsten sizes, you will need a gas lens for each size tungsten you use.

There are still many scenarios where a standard collet body will work best, especially in tight areas where you need a small diameter cup, or a long reach small diameter cup. However, for about 90% of what I do, I am using a gas lens. They are normally quite a bit fatter than the standard collet style cups are, and that's the biggest disadvantage

Instances where a gas lens simply wont work are usually rare. Normally I stick the tungsten out a bit further and that allows me all the room I need to make a good weld, but there have been situations where I can't use the gas lens, and I'm stuck running a XL, or XXL, or XXXL cup for a standard collet body. Its best to have both just in case. You never know what you're going to run into, and its good to have an array of choices.

The Welding Process

For most welding applications DCEN or DC electrode NEGATIVE is going to be your main process. This covers steel, stainless, chromoly, brass, nickel alloys, titanium, etc. AC will be used for aluminum mostly. Other metals like magnesium also need AC but for the most part you will be using it for aluminum.

DCEP is a setting that isn't found as much on newer machines. DC electrode Positive is pretty much only used to ball up tungsten on older machines for doing AC welding.

A lot of welders like to set the machine to DCEP and blast an arc on a piece of copper and this would cause the end of the tungsten to ball up, which worked well with older transformer machines.

On the older transformer machines, you are supposed to run pure tungsten electrodes, and with pure tungsten, if you grind it to a point, it wears away fast, and could contaminate your weld.

If your machine has DINS connectors, you can usually still switch them over for DCEP, that is, as long as the gas doesn't run through the DINS connector. Most will have a separate output for gas on the front of the machine. So, you can swap the leads, and the gas will still flow. But if your gas runs through the DINS connector, for example like on a miller maxtar 150, with the smaller dins connectors, you won't be able to swap them as your gas won't flow. If you can even achieve an arc, it will end up contaminating the tungsten. BUT. on a miller Maxstar 150, you can't weld AC anyway, so there really isn't a need to worry about it.

Besides. These days, tungsten electrodes have come a long way. It used to be that you would need to have two different types of tungsten electrodes on hand. You would have your red tipped 2% Thoriated tungstens for all your DC processes, and your Green Tipped Pure Tungsten for all your AC welding.

Nowadays they alloy the tungsten with other materials and they last longer and work well for all processes. I do still like my thoriated tungstens, and I still do use them the most, but I also am mostly only doing dc welding. My invertig is a DC only machine, and frankly, I'm not a huge fan of welding aluminum, (which I will touch on later) although, lately I've been thinking maybe its time to upgrade the invertig and get one that does aluminum just in case I need it. Its always good to have the option there.

If I had a need for welding aluminum often, I would probably switch my tungstens to lanthanated 2%, or even E3 tungsten electrodes. Both are good on the inverter machines and tend to last a long time between sharpenings.

So even on AC, with modern machines, there isn't a whole lot of reason to use a DCEP to ball your tungstens. If you have an old school transformer machine, you might like the way it welds better, But Inverter machines are square wave, which means the arc is much more stable and welds more like a DC in some respects, so you aren't really burning through tungstens like with a transformer machine.

So, you can just use your one type of tungsten, sharpen it the same way for ac and for dc, and not worry about it.

The main exception, and the only time you really need to do anything different is if you are using old style pure tungsten electrodes, on and old-style transformer-based welding machine where you will want to ball up the end for welding aluminum

To my knowledge, there isn't really a suitable welding application where you would use DCEP as the main mode of arc transfer for TIG. The arc is jumping off the base material to the tungsten, so all the arc force is pushing up into the torch, so even if you were to get a suitable arc, the arc force would prevent you from having any sort of penetration, and you would run through tungstens like nobody's business.

DCEP is mostly used in MIG welding. In mig welding, this is the most common polarity you would use, but a mig welder uses a completely different type of power supply, so with TIG welding, there really is no application for it.

Tungsten Electrodes

There are several options available to TIG welders for tungsten electrodes. The type of tungsten will vary depending on your application and welding machine. If you are using an old transformer machine, you'll want to use 2% thoriated tungsten for all your dc applications, and most recommend green pure tungsten electrodes for AC welding.

Newer inverter-based machines don't need the pure tungsten, and while some people stick to 2% thoriated tungsten electrodes, I've had the best luck all around on inverter machines with 2% lanthanated. it tends to work better on AC than the thoriated does in my opinion. But I still like 2% thoriated best for all around every day use. It's all a matter of personal preference, But I would suggest trying a few different types out and see what works best for your welding technique.

As far as diameter, 3/32 seems to be common for most people as an all-around tungsten size. I like 3/32, and I do use them from time to time, but my personal preference has been to use a 1/8" tungsten ground to a sharp point.

There are situations where smaller diameter tungsten is better for keeping the arc smaller and more focused, But I've found by sharpening a 1/8" tungsten to a fine point, I can do about 90% of my work and get top notch results. Very thin materials I will go 1/16" and I even have some as thin as .040. however, as I said 90% of the time it will be a 1/8" 2% thoriated tungsten ground to a sharp point in my torch.

When you sharpen your tungsten, there are about a billion different ways to approach it. Everyone has their own way that works best for them, but from my experience, there are some do's and don'ts.

Things to keep in mind are that we are dealing with electricity jumping off an electrode in an arc to a grounded base material. So, you want to keep the electrons flowing as freely as possible across that electrode. Grinding it improperly can cause the arc to wander and can make it difficult to start the arc in some cases. So, you'll want to have as smooth of a grind as possible, and you'll want to pay attention to the direction the scratches go in.

I personally like to use a belt sander to grind my tungsten. It gives me the finest grind and its quick and easy to position and get a good grind angle with a drill. Some people use a bench grinder, some people use a 4 ½" wheel on an angle grinder. It is recommended whatever you do, make your grind wheel that you use for tungsten exclusive to just tungsten grinding. You don't want to contaminate the wheel with aluminum or whatever and imbed chunks of an unwanted material into the tungsten.

The textbook correct, and accepted method is to always grind the tip so that the scratches are made parallel to the length of the tungsten, and to always sharpen it on the side of the stone that grinds towards the tip.

The thought process behind this is that if you grind INTO the tungsten, you potentially could have some grinding wheel debris get lodged into the material, which can contaminate your weld. Does it ever happen? I'm sure it has to somebody, but I've never noticed it.

To be honest, I always grind my tungsten with the wheel pushing the sparks into the electrode. There are a few reasons. First, is if I am using a bench grinder, I am grinding on the top of the wheel. Its easier to see what I'm doing, and I feel I have more control over the shape of the tip. This allows me to get that perfect taper, and the right sharpness that I like.

In order to grind with the sparks shooting away from the electrode on a bench grinder, you would have to be grinding on the bottom of the wheel and pushing upward against the wheel. I find this to be unnatural, and its difficult to see what the hell you're doing. Matter of fact I don't know how anyone could sharpen a tungsten this way.

It really does make a difference how you grind the tungsten. It may be a very subtle difference, but it's noticeable when you have a lot of time under the hood.

This is why I like to sharpen my tungstens to a fine point. I feel that the arc starts easier, it's more focused, and it doesn't wander as much.

I can't tell you how many times I've gone into a welding booth at a shop to use the tig, and the tungsten that's in the collet is all boogered up, And the pre-sharpened tungstens they keep around look like they were sharpened by Ray Charles. Absolutely zero attention paid to how they ground the tungsten.

I also notice with these tungsten electrodes that are poorly sharpened, for some reason I seem to dip them 10 times as often as I would with a clean tungsten. The arc wanders a bit, and the shape of the weld is sometimes affected in extreme situations.

The golden rule to sharpening your tungsten electrodes, however, is to always sharpen it such as to keep the scratches going parallel to the electrode. And if you think about what we talked about earlier with the electrons jumping off the electrode, this makes perfect sense.

If you grind your tungsten so that the scratches are perpendicular to the length of the tungsten, those electrons are going to have somewhat of a swirling effect as they come off the tip of the electrode. This creates a turbulent arc. It won't start as easily, and once you get it to start, you'll notice that sometimes the arc doesn't even come off the tip of the tungsten. Sometimes it's coming off the side of the taper.

When you have an arc like that, that's unpredictable to the point where you can't even rely on it to jump off the tip of the electrode, how can you be confident that it's going to produce a good-looking weld? You can't.

For the most part in my experience, one of the biggest causes for the weld to look like caca is a wandering arc.

Making sure your tungsten is ground nicely and is kept clean is vital to keeping your arc stable while you are welding. You want your welds to look like a stack of dimes when they are finished. The only way to achieve this is to have a very stable arc, and to keep it focused exactly where you want it, and where you want it only.

For that reason, I keep my tips ground pretty sharp. The only exception to this, is on jobs where I need to weld at a high amperage. Say anything over 175-200 amps.

The problem with sharpening your electrodes to a needle point at these levels, is that the arc will actually melt the very tip of the tungsten electrode, and it is possible for the arc force to blow that molten tungsten down into the weld pool.

This is an undesired effect for sure. Tungsten is a brittle metal. Very brittle. So brittle in fact if you take your electrode and tap it with a hammer it will break. It will crack lengthwise all the way down the electrode, and basically render that piece of tungsten unusable.

So, if you get a big glob of that tungsten melted down into your weld puddle, you are putting a brittle material that's prone to cracking into your weld, guess what happens to your bead?

It becomes prone to cracking. Does this mean it will DEFINITELY happen? Absolutely not. And under most circumstances, it would probably be ok. However, if you are ever doing any sort of code work, such as ASME pressure vessel and pipe welding, some of the jobs welded to that code will need to be X-rayed, or Ultrasound tested for impurities. If you have an inclusion of tungsten in your arc start, you may fail the x ray. Which means they will make you grind that weld out and redo it until it passes.

So, you'll want to grind the tungsten to a sharp point as usual when you're running higher amps, and then just take the tip and grind a very small flat spot into it. This will minimize the chances of you contaminating your weld and getting shit canned on a code job.

Welding is about best practices, so learn the "by the book" best way to do things, and then adapt. Figure out what works for you and where It's safe to skimp and cut corners, and where it's going to get you into trouble.

Work Area

The next most important part of getting a nice weld is having your work setup and positioned in a stable comfortable position. Having a flat metal table to work on is very helpful for TIG welding. You need to be able to keep the part from moving around, and you need to be able to keep your arms and body still enough to keep the weld bead steady without dipping the tungsten. So, it's important to spend some time setting up, planning your weld sequences, and figure out how to position your piece, and how to position your body to most comfortably make the weld.

This is the single biggest contributor to ugly welds. You can have perfect technique but if your wrist tenses up during a weld because you aren't comfortable, it's going to be noticeable when you dip the tungsten and the weld gets sloppy. So, spend some time getting comfortable and getting your workpiece setup and held down, and get your body positioned in a way where you are supported, but your arms are free to do what you need them to do without affecting your stability.

You do not need to necessarily clamp your piece to the table, but it's very helpful to have some weight on it. Sometimes if you are welding something just sitting on the table, the arc will make the part dance around on the table. It happens especially for me if I'm welding a baseplate onto a post, and I don't have anything anchoring the baseplate.

I'll start, and the part starts vibrating and rotating and it messes with your weld. Sometimes it will cause you to dip the tungsten into the weld puddle, sometimes it just moves and rotates to a position where you're now uncomfortable and can't weld. Either way, it's not ideal, and for perfect welds, we need as may conditions to be ideal as possible.

So, it's good to have some clamps or some weights to hold your parts down. investing in the tools, even if you must make them, in the long run will make your work come out better and will make you more money as a welder. You can use clamps, or you can fashion a device known as a third hand. Basically, it's just a weight that balances on 2 feet, and has a single point that will apply pressure to your part. You can make one of these in about 5 minutes with scraps laying around your shop in many cases.

Using one of these will often keep your part from moving around while you weld, and it also has a couple added bonuses. It can be used as a remote grounding point.

Many people will ground their table and place their work on the table, then just weld away. Well this is ok in most cases, but there are sometimes instances where there might be an arc between the part and the table, and if you are working with something that's already finished, like brushed stainless. This could completely destroy the part you are welding, causing you to spend lots of time trying to salvage the finish and blend it out.

The third hand, can be used as a place to ground your workpiece in a very specific spot. For example, you can use it to hold down a piece that is finished, but eventually has another piece welded to it. Place the third had where the next piece will be welded to and connect the ground cable directly to the third hand. Even if it does arc, which it likely won't, it will be covered up by the next piece.

One of the shops I worked at, the guys who ran the place had never heard of or seen a third hand before, and they saw me in the back of the shop making this little contraption. Basically, all it was a piece of stainless steel solid bar stock about 2.5" in diameter and about 5" long. So, it had some weight to it. On one end, I welded 2 ½" L pins that we had laying around to use as standoffs for a railing. They were made of ½" stainless solid bae. The two pieces were welded in such a way as the outer corner of the L shape would act as a pivot point. I welded them to one end of the 2.5" solid bar, and on the other end, I welded a third L bracket that came basically straight out and hooked downward. I ground the short end of the L to a blunt point, and this was the pressure applying point.

Basically, it pivoted off the 2 L pieces on the bottom applying the weight of the solid stainless bar to the single blunt pointed tip.

The guy who ran the shop looked at me like, what is this thing? I was like "it's a third hand. its for holding things down." And he literally gave me this look like I was a complete moron and laughed at me. Within weeks he saw how much I had used it and how useful it was, and before I knew it, he was using it too, and it became one of the most useful tools in the shop.

But this was a shop where everything had to be brand new and nice looking professional brand name tools, or it was basically junk. I've gone most of my welding career using tools that I personally made, and they always end up being my favorites. I've made hammers, chisels., my own chipping hammers, pry bars, clamps, you name it. That's sort of the point of being a welder. you make things.

Now don't get me wrong, you can't make everything, and even if you could, sometimes its just not worth your time to do so. But little things that you can make out of scrap, can usually help you out quite a bit. So, they are good to have around. A third hand, however is almost mandatory in my opinion. It makes your life so much easier. I liked the one I made at that shop so much I wish I had brought it with me when I left.

So, another important part to making beautiful welds is being able to clearly see your work. Using a crappy helmet, or a dirty lens, or whatever simply isn't going to cut it. If you really want to take your work to the next level, you really need to be able to see with some clarity.

I recommend getting a good auto darkening helmet with a good size viewing area. I personally use a miller digital infinity, and aside from the design of the headgear, it's been a pretty solid helmet. I've had it for about 3 years now, which doesn't sound like a long time, but for someone who welds all day every day, who is bringing it out on the road all the time, and dropping it every now and then, that's actually quite a while. I will say there was a point in time where I HATED this helmet. I kept having trouble with getting flashed. I couldn't figure out why. I kept changing the battery on it and it wasn't helping.

Then I took a second look and realized there are TWO batteries in two separate battery slots. Yeah. idiot move of me, but once I changed both batteries I haven't had a complaint since.

I have also used some of the cheap Chinese auto darkening helmets, like the ones harbor freight sells. In many cases they are enough to get you by, and they do what they say they do. And realistically… they aren't terrible for $40. But you will notice a huge difference in clarity between the two helmets, and the more clearly you can see, the better you can weld.

So, it really is worth the money to spend on a good helmet. You only get one set of eyes, and you can get something good for about $300 or less, and that's money you won't have to spend for quite some time. I will recommend though, if you get a miller digital infinity, look at replacing the head gear for the style they used in the older digital elite. The one in the infinity is comfortable, but it's a pain in the ass. Its not worth the hassle in my opinion.

As for the rest of your gear, I recommend wearing gloves, although I will admit I myself do not wear gloves 90% of the time. As a disclaimer, I have been burned and shocked by not wearing gloves, but it's rare, and for me personally, I like the feel I have over the filler rod with no glove.

Feeding with a glove is a bit less precise, and if I am trying to make very perfect looking welds, I want to have as much control over my filler rod as possible. If I am welding higher amps, where the work I'm doing is hot, or if it's out of position, then I will use gloves. But if I can make it happen where I'm not getting burned, and can still be comfortable in my positioning, I'm going no gloves.

You can also get yourself a set of gloves that fits your hands well enough, and then use a propane torch to shrink the leather, so that the gloves fit your hands perfectly. This will give you both the extra control you are looking for, and the safety that you really do need.

Again, I recommend you do not weld gloveless as you can get injured, and it's just not good practice for the work environment, but if you're a hobbyist you might find you like it better.

Just be careful because you can get zapped. if you accidentally ground the filler wire and the arc jumps off the tungsten to the filler wire. It will burn your hand faster than you know what hit you, so you need to be careful. This has happened to me on a few occasions where I was working in a Tight area surrounded by grounded metal with a fresh piece of filler wire. Now I like to shorten my filler rod in situations like that.

One time I was installing a stainless-steel spiral staircase on top of the roof of a mansion out here in Connecticut, and the staircase had a floating baseplate that was to be hidden underneath the floating stone floor. The floor tiles sat on these risers, and below them was some type of galvanized decking to protect the roof.

Well I had to weld some stabilizer bars to the center post, that would eventually be screwed into the walls to support the stairs, and this was before the flooring went down, so I was laying on this galvanized decking, trying to get a weld on the center post, and I didn't shorten my filler rod. I was laying down on my side, and the weld was about 3" off the ground, and the end of my rod ended up touching the galvanized sheet. Soon as I tried to start my arc, the arc jumped to the filler rod, and basically the whole filler got super hot and burned a line into my skin. Of course, I was wearing no gloves like a true professional </sarcasm> so now I don't do that anymore.

I always cut my filler rods down if I am working in a tight space, and I try to put on gloves whenever possible. Sometimes its nice to have that extra feel, but sometimes it's not worth it.

So. Next topic. Shielding gas. When it comes to your shielding gas, 100% argon is going to be your gas 99% of the time. If you do a lot of aluminum, you may want to consider using an argon/helium mixture as helium will allow the puddle to get hotter, meaning better penetration and ability to weld thicker materials.

But for our purposes in this book, we will assume 100% argon at a flow rate of around 15-20cfh. This is a pretty good all-around flow rate for in the shop use. And in most situations, 100% argon is going to be the best gas. For pretty much any sort of tig welding. I like to run a little more gas if I'm using a larger cup, which I usually do, so my machines tend to stay around 20. I typically use a number 8 cup for almost everything.

The only time I will really feel the need to change out my cup is if I need to weld in a very restrictive area, and my cup is just too big, OR, if I need even more gas coverage, and want to switch to a bigger cup.

I have played around with using cups as small as a number 4 with a gas lens, and I do like to use those from time to time. Mostly if I am looking to make a very small weld.

Fun fact: the number of the cup is actually the measurement of the cup in 16ths of an inch. So, a number 4 measures 4/16 or ¼" a number 5 is 5/16" a number 8 is 8/16 or ½", and so on.

Many people these days are running these gigantic cups for just about everything. I like them, they have their applications, but you should be able to get perfect coverage with any sized cup. I know on stainless steel, I can get absolute perfect color with any size cup. It just takes a certain feel. You know the metal is the right temperature because its flowing right. You know, and once you are used to it, your travel speed plays a big part in coloration. So, I think one of the downsides to these monster cups I see everywhere is that it makes people's welds have perfect color regardless of technique. There's just so much gas coverage that there's like no way to have too much heat. The weld is cooling much more efficiently than with a smaller weld.

With materials such as stainless steel, weld color is an indication of the heat you are using being too high or too low. So, if your cup covers for your bad technique, how do you ever learn to dial in your settings with a smaller cup? Just an observation, and you can take it with a grain of salt. Many people using these monster cups still lay down some beautiful welds.

I had a customer drop off a manifold to be welded that he paid $1,000 for. It was a turbo header for a Honda, and the company he had gotten it from was pretty well known online for their headers for all makes and models. Well whoever welded it did a shit job, and didn't weld with enough heat, and 2 of the Pipes cracked at the weld, and one of the pipes ha complete weld failure in at least 2 joints.

The customer said that he was on the highway, he was ripping on it, and it blew apart and he heard the piece dinging down the highway.

The welder did not use nearly enough heat to penetrate, and probably also had a giant cup where he was unable to see his heat input because the cup was covering up for his poor technique.

On stainless steel, a silver or gold (straw) color is ideal. With these giant cups you are almost guaranteed to have a silver or straw-colored weld every time, because there is just so much gas coverage.

So, if you're inexperienced, you could easily weld something cold, and not know it if you aren't exactly sure what to look for. Which I think was a large part of what happened here. I think this guy just learned how to weld in his spare time and decided to open a business fabricating performance parts.

Now. some of you are probably saying to yourselves "well that's what he gets for driving a Honda" and frankly, to some extent I agree. But if you pay a grand for a manifold, it ought to at least be welded properly.

The welds were clearly VERY cold, and any well-trained welder would have been able to spot this very easily. I even posted a photo of the broken header, and tagged the company who welded it on Instagram, hoping maybe they would make good on their workmanship, and got no response. The owner of the car said he called and called trying to get a replacement, and they told him there was nothing they could do.

I'm assuming this was an actual problem for them because if it were me, and I had someone call up and say their header broke, what's it takes for time to fix it? I spent a few hours taking off the shelf pipe sections and cutting and fitting them properly. They already have the pieces cut, or so you would assume. Eat the loss, send out a new header, or fix the old one. Its just good business, especially when its clearly their fault.

So, you need to be observant about how much heat you use. Just because a weld looks nice, does not mean it's structurally sound. Luckily this guy who welded the part sticks to exhaust and turbo systems.

Now imagine if he had welded something structural, such as say a balcony frame, or a staircase, and all of his welds had zero penetration. Somebody could easily be killed in a situation like that.

Proper Fit-up

So now that we have talked about machine setup, let's talk about setting up your work. Let's talk about how your pieces should fit, and what to do if they don't. This is another big deal when it comes to getting a good looking and strong weld.

There are situations where you'll want your pieces to fit tightly together, and there are other situations where that may not be the best choice for the application. There are times when a slight gap helps with penetration, But, too much gap can cause major problems as well. Everything is a trade off, and there's many ways to skin a cat, as they say. So, you'll want to know several different methods of fitting together your parts so that you know what to do when you run across these various situations.

When you are fitting your pieces together, if it's not stated on the print, (if you are working from one) it's important to consider the application and how much stress the joint will see. If the joint is structural and will have stress applied to it, it is wise to bevel your pieces to allow the filler rod and puddle to flow in deeper into the weld joint.

On thinner materials, instead of beveling, you may want to just leave a small even gap between the pieces. However, that doesn't mean you have to do either/or. Sometimes I will do both on the same joint. A piece of filler rod works well as a spacer. To keep your gaps even.

This is a good technique for situations where you want full penetration on your welds, as welding with a gap will allow the puddle to flow through to the back side of the weld joint. Applications that see lots of stress or internal pressure, such as tanks, pipes, or pressure vessels are often welded this way to achieve full penetration. The trade term is open root welding.

The root is basically the groove that is formed by the bevel, so when you run what's called a root pass, you are bridging that gap, and tying those two bevels together. A root pass is followed by one or two more passes called fill and cap passes depending on the weld procedure.

Open root welding is very useful and creates very strong joints, however, you may be welding a piece that doesn't necessarily have to be very strong but needs to be precise. This is where fit-up becomes a bit more critical.

Whenever a joint is welded, there will be some degree of shrinkage. You are adding filler material while the metal is expanded and hot, so as the weld cools, it's going to contract a bit and pull in on itself. The amount of shrinkage is going to depend on material, amperage, and filler material. For this reason, it's important to pay attention to how the metal is pulling, as it can fight you in the long run if you aren't careful.

Often times you will use open root welding on pipe and pressure vessel type applications, but every now and then, if I have something structural, and I just really want to make sure its penetrated deep, I will run open root. Especially on thicker materials. ½" and larger thickness. Many times, I will bevel the pieces AND leave a small gap. This ensures that the weld burns deep enough into the groove that I can get 100% weld penetration.

Shrinkage however is the biggest issue with doing things this way. Its usually not a huge deal when you are welding pipe and pressure vessels, because its's accounted for when they cut the pipes to length, and you are welding in a complete circle around the pipe usually. Which means it will pull in on itself evenly around, and you really won't notice much distortion.

Where it becomes an issue is when you are doing fabrication and welding structural pieces that need to be located and stay located. Since the weld amperage affects the amount of pull, you can't always predict how much it's going to move. So, in these types of cases, you'll want to keep a tight fit on your parts and use an adequate sized bead to get your strength.

So, let's say were talking about welding a square frame out of 1.5" square 308 stainless steel tubing, and we want to keep it perfectly square. Let's say the frame is 30x30 and all 4 corners have 45* miter cuts.

Because of how the metal pulls when you weld it, it's important to have a very Tight fit up. Any gap in the material will allow more room for contraction when the weld cools, so even if you clamp your pieces at 45, they will contract, and spring closed a degree or so when you unclamp them.

This is bad if you are looking to keep things square, so you want to make sure your cuts are good, and if need be get them as close as possible with a grinder.

There will be situations where you can't have a perfect fit up, and you must deal with what you have. Your results in these situations won't be perfect, but you can get by if you are smart about how you tack the metal.

The important thing is to clamp and position the materials to your table, and then tack joint on opposite corners starting with the corner that will pull least (usually outer corner) you can then sort of fill the gap with a few tacks, you may need to spend some time bridging the gap, but the idea is to lock the gap in place so that when you go to make your full weld pass, the joint can't close in on itself too much. There will always still be that small amount of pull, but if you lock in your gaps with tacks, the movement will be minimum

Now if you tried to weld that same gap, especially with stainless, and you started at the closed end, welding towards the end of the tubing with the gap, by the time you got to the end, the pull would be so strong, it would most likely completely close the gap by itself. Which leaves you with parts that are fully welded at an incorrect angle.

On thicker materials, you may need to bevel, or use gaps to get a full penetrating weld. It will depend on the application, and your machine's capability. I like to at least cut the edge on pretty much everything. If you are using a perfect fit-up with crisp edges, sometimes it is easy to lose track of where you're welding. So just sanding the edge off or adding a slight bevel can be helpful in keeping track of where you are going.

On larger materials, where a larger bevel is used, you can walk the cup, and lay the wire right into the bevel. This is a pipe welder's technique and speeds up weld time tremendously, it does take a little practice however.

Walking the cup is a technique in tig welding, also somewhat can be done in mig welding, where you rest the edge of the cup on the bevel or weld joint. And using a loose figure 8 motion in your wrist, you roll the cup back and forth across the weld joint. Sort of the same way you roll a 55-gallon drum across the shop.

To roll a 55-gallon drum you tip it back so its standing on the edge, then roll it to the left a little, turn the barrel, roll it to the right, then turn the barrel back, and roll it back to the left, then turn it again and roll it back to the right, etc.

To walk the cup, you want to basically achieve that same motion with your cup, and its difficult to explain, but I would call it a loose wristed figure 8. You grab the handle of your tig torch, and usually wrap it around your wrist once so the tension from the power lead isn't pulling on the torch.

And you'll roll your wrist one way, then sort of angle the cup the opposite direction, and roll back the opposite direction.

It sounds complicated, but truthfully, its incredibly easy. Its honestly easier than getting a clean weld by just dabbing in my opinion. But back to fit-up.

For most situations, a Tight fit with a mild bevel will produce satisfactory results. If you are welding at the correct temperature, with correct technique the puddle will penetrate, and it will be a strong weld.

One of the most time-consuming parts of metal fabrication is the actual fabrication. The cutting and fitting of parts, but if you spend the extra time to make sure you have a Tight fit, in the long run your welds will be much nicer. Filling gaps makes the weld look sloppy, so you want to have a pretty Tight even gap, so the weld bead lays in evenly from start to finish.

This will get you where you want to be on about 99% of your applications.

However sometimes just a small gap, the thickness of your welding wire makes the weld penetrate much more and gives you a stronger weld.

It can be tricky and sometimes you need to brace or gusset your pieces when you weld like this, as the gap can sort of flex however it wants to. So, you can end up with an undesirable bend in the joint. An example of this would be on a butt joint of two pieces of flat stock.

The pieces will tend to bend towards the side you are welding on when they warp, so if you leave a gap, and there is nothing supporting the edges, it could bend even more than if you had but the pieces directly up to one another.,

It's going to depend on the heat and the thickness of the material, an honestly the shape of the part as well.

On some tanks I built, the sections were fabricated out of 3/8" thick 316 stainless. The outer diameter of the tank was around 11'6" and so there is no stock pipe you could buy for this. So, they had to custom order the material in extra long lengths, and then put them through a roller. Once they were put through the roller, they welded on some braces called strong backs.

These strong backs were thick pieces that got tacked on perpendicular to the joint, so that when we welded the seam, the tank stayed round at the seam.

Without those strong backs, there would have been some major flex and the rest of the tank would be round, but right at that weld seam it would cave in like a V where the weld was. The strong backs were thick enough to prevent movement of the material during welding, but also allowed room to get the welder in there and get a full penetrating weld.

Weld Prep Procedures

Another huge part of proper welding technique is how the materials are prepped. Every material has its own sort of nuances as far as prep work goes, but they all have the same basic things in common, clean, clean, clean. you can't have any grease, or oils, or rust, or dirt, or paint, or grime, or residue on the metal. This is pretty much true for all materials.

When you strike an arc, and begin to melt the weld puddle, you will be introducing those contaminants into the weld if things are not clean.

Needless to say, this is not good and will lead to dirty, sooty welds with porosity, and undercut and pinholes and all that crap. So, if you want to make nice welds, you need to spend the time to make sure the material is fit properly and prepped properly.

A lot of people tend to skimp out on this, but it's really the most important part. I always hear of people who think it's ok to just burn through paint, dirt, grease, grime, etc. and as long as they can strike and maintain an arc, they don't need to worry about getting things clean.

Even some shops would give me crap about how much time I spent prepping things. I would be in my booth wiping things down, and trying to get them ready to weld, and someone would come over and say "you don't have to go crazy over here man we just need to get this stuff welded"

I hate that crap. Because basically what they are saying is they don't care what you do, or how the quality is, just get it done as fast as humanly possible, and if it means not doing it correctly, then don't do it correctly.

But on the other hand, as soon as something goes out screwed up, they are the first people to come chew your ass out when the customer calls up angry that their parts broke or whatever.

So, I don't pay much attention to people who want to rush me. The more time you spend prepping, the better your work will come out. This is true not only for welding but for most trades.

For carbon steel, you must work against having a mill scale. The mill scale is a layer of scale caused by the manufacturing process, where steel ingots are heated to orange hot, and ran through rollers to form either sheet, plate, tube, angle, or whichever type of stock you are working with.

Typically speaking, you cannot safely weld to the mill scale. However, I have done it several times in my time as an architectural welder. We had requirements for the job to leave the mill scale and have no grind marks anywhere.

This was the look the architect was going for. I never liked doing it, because the results are unpredictable, However, I found that you can get a satisfactory weld provided it is not a critical application, but there is a certain technique which I will discuss later in this book.

For most situations, you will grind the mill scale off the material with an angle grinder, and your weld will be made on clean shiny steel. This is going to give you the cleanest weld, and it's going to allow the puddle to flow best.

The best ways to do this are to use a flap wheel or a sanding disc. You can obviously use a grinding wheel, but this leaves rough scratches in the material, and if it's gong to get painted, you'll most likely see those scratches through the paint. If you're going to take the time to make your welds look good, you should also take the time to make sure your paint looks good too.

So many times, if the mill scale is very thick, I will use a grind wheel to get most of it off, then go back over it with a flap wheel or a sanding disc to blend out all those scratches, and make it look smooth.

After you grind all the mill scale off, it's important to get all the grease, dust, dirt, grime, and other crap off the material. The grinder will leave some dust, and it can interfere with your welding. Plus, if you are wearing gloves that you've used before, they will have dirt and grease and what not on them, so its important to get rid of all that.

Wipe everything down with a good solvent like acetone, or lacquer thinner. Prep ALL is another good choice, but my preferred cleaning solvent is lacquer thinner. It works well, and it evaporates leaving no residue. I usually have a methodic approach to cleaning my materials. It is however not friendly for plastics, so watch the rags you use. I can't tell you how many times I'd grab a rag and some lacquer thinner to wipe a part down, only to put the rag down somewhere and set my helmet down on it a few minutes later. The thinner melts the lenses and the plastic, so you need to replace the lens. Those can add up in price, especially if you have something like a miller digital elite or digital infinity, that uses miller specific proprietary lenses.

Typically speaking I am a bit methodical about how I do my prep work, and fabrication when making things.

I will first do all my cutting. I get every single piece I need to do the job cut to length, once I have everything cut and ready to go, I will deburr my edges with a grinder, and usually clean the edges of mill scale at the same time. A belt sander is great for deburring your parts. It will leave the cleanest finish, so your parts look that much better once you have it all together.

Once I have everything ground, deburred, and the mill scale removed, I will then wipe the entire part down with a solvent, cleaning off any residue or dirt, oil, etc. typically just a shop rag and some lacquer thinner does the job. And occasionally I'll have to do it twice, if the material is really dirty and dusty. It's worth getting it all clean in my opinion, as the results will be much better.

So once that's done, I will fit my parts together, and clamp them if necessary. Sometimes you don't have to, and its ok to just tack it up while holding it together with your hand.

Once everything's tacked, for good measure I will wipe everything one more time before final welding. You will spend extra time doing these steps, but it is worth it in the end when your carbon steel welds come out looking just as good as stainless-steel welds.

Typically speaking carbon steel isn't known for being a very "pretty" looking material when you weld it, however with the right prep, I disagree. You can get some very nice colors out of a carbon steel weld if you take the time and make sure the material and weld area are clean clean clean.

For stainless steel the procedure is very similar, except there is no mill scale to grind. I will do my cuts, do all my finish work before hand, weather its polishing or sanding. Wipe everything down with lacquer thinner, and then I will go ahead and tack my parts together.

When everything's tacked sometimes I'll go through and just clean with a brush to make sure there's no soot, and wipe again with a solvent.

Sometimes when you tack stainless, you'll get a little soot around the tack weld, and when you go to weld, the puddle will pick all that crud up, and you'll end up with a bad weld in that area. So, if you see the black soot, just go ahead and clean it off. It only takes a second or two.

Another one of the biggest keys to working with stainless is that you need to do it pretty much away from any carbon steel work. You can't place stainless steel parts on top of a carbon steel table, you can't grind carbon steel next to stainless steel, none of it.

The reason being is that any dust or debris from steel can imbed itself in the stainless, and it will eventually rust. Even if you took a block of steel and just rub it on the stainless, it the stainless will pick up some of the steel and eventually you will start to see rust develop.

Most people pay for stainless because they want a low maintenance finish that won't rust. so, when you aren't careful and imbed their stuff with carbon steel dust, and it rusts in two months, you're going to have a customer in your face wondering why these parts are rusting when you said they were stainless.

And that's another thing. Stainless does not mean literally stainless. It can and will corrode all on its own. One of the shops I worked at did some railings for a customer whose house was on the ocean. Like their back yard was a private beach on the Atlantic Ocean. So, they had ocean mist blowing at these railings all day every day, and the salt corroded them pretty quickly, and they began to look like crap.

They specifically ordered 316 stainless because they thought that 316 was somehow impervious to corrosion. And within weeks the salt began to corrode the stainless, and it looked like crap. This had nothing to do with us not being careful about having carbon steel around, but it's to let you know that stainless is not magic. It can and will corrode for various reasons. And the most basic of them is carbon steel contamination.

So, if you're working with stainless in a shop where you usually work with steel, you need to take a blow gun, and blow all the steel dust off your tables, you need to wipe everything down with lacquer thinner and make sure there's no steel dust laying around. If it's going on a stainless-steel railing inside an office, it probably won't rust, but if it's going on a railing at a beach house, or if it's going on a tank that gets washed out with chemicals, chances are that rust is going to come out. So, you must be very careful about how you handle stainless. The last thing you want to hear is that your customer at the brewery had to dump 1,000 gallons of beer because the inside of their kettles started rusting because you shot steel sparks at it

Other than that, stainless steel is a very easy metal to prep and weld. Just clean, clean clean, and clean it one more time. Another tip for stainless steel, is if you use sharpie to mark things, you can tack things, but before you weld clean the sharpie off with a rag and some acetone. The heat will actually etch the marker into the metal and it won't come off without some real effort.

Aluminum is a little trickier. Aluminum has a layer of oxide on it that pretty much begins forming immediately after you remove it. So, you must clean it just before you start welding.

The problem is the oxide layer melts at a much higher temperature than the aluminum underneath, so if you don't clean it off, you'll end up melting the metal below the oxide, but you can't get the filler rod to stick to the puddle because the oxide isn't melted, and you end up with little gobs of aluminum that usually don't stick to anything, and a whole lot of soot.. And if you are not careful, can actually ruin the material. If you end up getting too much soot in there that you can't clean it out with a brush, it's going to be difficult to make a good weld.

So, to get that oxide layer off, my personal method is I will scrub it with a wire brush and then scrub it with a scotch brite pad, then wipe the whole thing down with acetone or lacquer thinner that's usually enough to get a clean weld.

I don't suggest grinding because you can embed crap into the aluminum which makes it much more difficult to weld. At the most, if you must, sand it using a flap wheel, a sanding disc, or a belt sander. But I would avoid using a grinder with a hard wheel on aluminum. Fiber resin discs and flap wheels are better choices.

Another trick you can use, and this is helpful if you are welding on dirty aluminum is to run a cleaning pass with your TIG torch before you start welding. You can do this by simply lighting an arc, but slightly too cold to actually weld, and you will just run over the joint just until the very edge of the metal starts to round over a little bit.

You aren't trying to fuse or weld the pieces yet, all you're trying to do use the cleaning action from your AC setting to break up the oxides and any other crap on the surface of the aluminum. If your machine as AC balance you can turn up the cleaning action before you give it a pass, and this will help.

This technique is also great if you are using a smaller 200 amp or less machine and trying to weld some big thick aluminum parts. It actually serves a second purpose, and that is it will preheat the metal for you, allowing you to weld it at a lower amperage, or at the very least it will allow you to get a puddle formed faster.

If you are welding thick aluminum parts, a clean pass won't hurt. Just be careful not to overdo it on the amperage. You aren't trying to form a puddle, you just want to barely melt the edge of the material so that it rounds over a little, and you can see shiny metal.

Brass preps just like steel or carbon. I usually just scuff the weld area with a scotch brite and wipe it with an acetone-soaked rag.

You will get to be familiar with each material in time, but the standard rules apply to all materials; Cleanliness is key. You really need to be careful not to introduce contaminants into the weld. Welding is a game of best practices. It's one of those things where you might get away with not cleaning your work with acetone 7 times out of 10, But that one time there will be an invisible layer of crud on the material, and it will be a critical joint both structurally and aesthetically, and you will have porosity and soot, and a ruined part. Trust me, it sucks. It's much easier in my opinion to spend the extra time to wipe your parts down a few times to make sure your weld comes out perfect the first time, because at the end of the day, if something gets screwed up, you're the one who looks like an idiot when your weld is full of porosity and holes.

So, spend the extra time and do it right. if you're slow but do a good job every time, your employer will want to keep you, but if you rush everything and skip steps and only 60% of your work comes out top notch, they won't want to keep you around very long. So, don't rush, it will show in your work, and will ultimately leave you frustrated in the unemployment line.

Even if they are on your back to work faster, don't skimp on quality. And don't be afraid to try things out, just understand that if you screw something up trying to find an easier way to do it, you need to be able to cover your tracks.

I often say being a good welder isn't always about doing everything perfect every single time, it's about knowing how to cover your tracks and fix your mistakes without anyone knowing there was a problem to begin with. The neater and cleaner you work, the easier that its.

It will be easier to cut apart a joint that's Tightly fit, as you can assume the center line of the joint much easier and get a more accurate cut. You will mess things up, it does happen, Stuff moves when you don't expect it to, shit happens as they say. So, learn to work efficiently so that if something does go wrong, it's not the end of the world and you can salvage and recover the pieces easily.

This is why it's important to have a method to your madness as they say. If you do thing according to a plan, or a routine, it will be much easier to retrace your steps backwards and fix things that you may have messed up on.

.

Tacking Your Work

So, one thing I didn't get to much into in earlier chapters is tack welding. When you fit two pieces together, you need to tack them obviously to hold them in place while you weld them. It's important to place many tacks at strategic points to reduce heat distortion (warping/pulling). Wherever you place a tack, place one exactly opposite to that. You want to have the tacks sort of fighting each other so that the piece can't really move in any direction.

When you make a tack, especially on stainless steel, the piece you tacked on, will pull towards the tack. So be sure to check your work after the first and second tacks to make sure everything is still where it's supposed to be, and then double and triple check before you fully weld.

There are a couple techniques for tack welding that you can use. A problem many people will find is if they can't clamp the piece, it can be difficult to hold down for tack welding. In most cases, if your fit-up is proper and you have no gap, weather the part is beveled or not you can fuse tack the parts together.

I typically use a lot more amperage for tacking because I want the tack to melt very quickly so the part doesn't get hot while I'm holding it.

You are only holding the pedal for a split second. basically, floor it, and let go in almost one motion. This creates a really quick blast of full power welding amperage, which is usually strong enough to melt the material in that one spot but does not get hot enough to actually warm the whole part up.

With a clean sharp tungsten, this should send a tack right into the joint. Do the same on the other side until the part is tacked and holding its own weight, then you can go back and add a couple tacks with filler rod and make your first tacks a little bigger.

When you add filler rod, you want to basically warm up the puddle, then dab the filler into the edge of the puddle. The puddle will take as much filler as it needs to build up the tack. Sometimes you need a dab or two to get it as big as you want, but you don't really want to feed rod into the puddle.

What happens when you put too much rod in the puddle, is that since you are just putting a tack in, the base material is still probably not much hotter than room temperature. There isn't enough heat to keep up with adding the filler. So, if you add too much filler, the rod will actually steal heat from the puddle, and you'll end up with an improperly fused tack. So just touch the rod to the edge of the puddle and the puddle will take what it needs.

Its good to do this whenever possible, because in most cases, simple fusion tacks aren't strong enough to keep parts together.

Especially if you're working with carbon steel. In some cases, the fusion tacks will be enough to hold it together, but any little bit of stress can cause them to break. so, I like to add a little bit of rod when I can for strength.

I have had fusion tacks on carbon steel break on me many times, and it's annoying when you need something to hold and it doesn't. it's worth the extra few seconds to just go ahead and put a few dabs in.

However, I should note that Stainless steel is much more friendly to fusion tacking, and fusion welding without filler rod in general, and there are instances where it's acceptable to use no filler at all. Kitchen equipment is usually just fused together with no filler. Typically, there isn't much stress being put on a weld on a stainless-steel sink, so in order to save time on grinding and finishing things, they fuse everything.

I've even done fusion welds on things that would seemingly be somewhat structural as an architectural welder. Railings most of the time are fused together instead of welded with filler, and when you have so many wed points on a railing, all those welds work to keep each other straight and strong, so it kind of works out.

For tacking aluminum, it's a similar procedure but its slower. When you start an arc on aluminum it will take a second to heat up, and you'll see different stages of the metal. First the arc starts, then the material starts to look a little dirty and cruddy, you keep working the arc back and forth from edge to edge, and you start to see the edge melt over a little, hold it a little more and you get a football shaped shiny silver puddle starting to form.

Keep holding it a little longer until it starts to wet out into more of a circle than a football shape, and then add your filler rod. You must be quick about it, because once you get to that circle shape, that means your puddle is up to temperature, and its penetrating, and if you hold the arc too long, it will take a dump on you and leave a big hole for you to look at. This sucks when you're dealing with machined aluminum parts that can't be repaired, so you should be careful and pay attention to what the puddle is doing.

When you are backing down off the tack, the same is true with all materials, but aluminum especially. Slowly circle the puddle around and around at the end of the weld, add a few extra dabs of filler to keep a crater from forming and slowly back off the amperage until the arc dies. Hold the gas over the weld until it cools a bit. It's a very similar process, but the material melts a little differently, so it takes a little bit longer, and you are looking for different signs. Otherwise it's not a very difficult material to work with. The biggest issue most beginners have with aluminum is overheating the work. This goes back to that pesky oxide layer on the material. This is another reason why its imperative to clean up the material properly.

Even if you are just trying to tack a piece of aluminum, that oxide will prevent you from ever getting the two pieces to fuse together. You will keep adding heat to the parts hoping they will fuse, but again, that oxide layer melts at more than three or four times the temperature that aluminum melts at.

So with beginners, they will be sitting there with the pedal to the metal just pumping crazy amps into this piece to try and get a tack, and they think that they just don't have enough heat, so they keep holding the heat on that spot, and then all of a sudden, they've raised the temperature of the part to the melting point, and a big chunk of the part melts and sags. At that point it's basically useless. Throw the part away.

But, if you clean that oxide layer off, aluminum only melts at like 1200 degrees, so it normally welds very nicely, and when its very clean you can do typical fuse tacks. But just as with steel, you'll want to fuse tack them just enough to keep your parts together until you can get some filler wire stuffed in there.

The nice thing about aluminum is it tends to stay put when you tack it. The real movement happens when you full weld pieces, because there will indefinitely be some shrinkage. But when you are just tacking things, in my experience, aluminum doesn't tend to pull too hard one way or another.

Carbon steel will definitely pull a little when you tack it, but its really not all that bad if you do small fusion tacks to start, then lock the pieces down so it can't warp when you go back and add more filler.

The biggest culprit of movement when tacking is stainless steel. The reason you always hear that stainless is such a hard material to weld really has nothing to do with actually running the bead. It's because it doesn't want to stay straight.

Even a quick tack will cause distortion on stainless steel, so it's important that you put as many tacks as you can fit on your workpiece to ensure it does not move.

Welding Technique

So, let's talk a bit about weld technique now. There are many things to be said about tig welding technique. Everyone is going to have a slightly different approach. I'm going to talk about what I do personally, and why it works for me. There are some people who may feel they have a better technique, and that's fine. If the result produces a good solid weld with no undercut and proper penetration, without overheating the material, that's all that matters. At least in my opinion.

So, I start by setting my machine. I'm a firm believer in using what I need for heat and no more. I have worked with many welders over the years who just set the machine as hot as they can and use the pedal to find the sweet spot. This approach works in some cases but does not give you the best-looking welds in my opinion. The reason being is you are heating the metal to a hotter temperature, which causes it to run thinner than normal, so your puddle isn't going to flow as consistently as it should. When you have a really "watery" puddle, for lack of a better term, its going to run in ways you probably aren't expecting. You'll never get the same consistency you want, and the welds usually end up looking pretty sloppy in most cases.

The other obvious issue with this approach is that you end up putting more heat into the material than you need to. This can lead to more warping, larger heat affected zones which on some materials can be a sign of the material weakening outside of the weld bead.

You can crystallize the material, which is what's happening when you get those overly gray burnt looking welds, especially on stainless and other nickel alloys. But it can happen on carbon steel too.

What happens is you end up overheating the area around the weld, which causes it to temper. When you temper material, it becomes slightly harder and more brittle, so the chances for it to crack and fail in the future are increased, but not guaranteed.

So, you want to spend some time getting to know the machine and the material and figure out what a good consistent setting is for your technique.

And what I mean by that is, it may not be very comfortable for you to weld really fast. Its better for the material, but maybe your hand just doesn't have the stability to run beads really quickly. Maybe you need to weld a little slower.

Or maybe you're like me and prefer to weld fast and hot. Either way, there are certain settings you'll have to have your machine set to, in order to get your desired result.

When it comes to making a weld, I believe in welding hot and fast. The key is to reduce heat input. When you weld very slowly, you are keeping the heat on the material longer, which allows the metal to absorb and hold onto more of the heat.

You can end up with burnt/crystallized material, or brittle joints. Welding with a higher amperage at a faster travel speed will allow the arc to bite deeper down into the material, giving you good penetration, while keeping overall heat input to a minimum. The part will still most definitely get hot, but you can look at your heat affect areas, which are the discolorations that appear around the weld, and tell the difference between one that was welded correctly, and an incorrectly made weld.

When a weld was made slowly and too cold, it will have a large and usually inconsistent heat affected area. It will be blotchy and usually starts off small and gets bigger and bigger as the weld progresses. Its more commonly like this on mig welded joints, where you have a cold start, and eventually build heat up to achieve proper penetration, But I've seen uneven heat affected zones on plenty of tig welds too.

A weld that was made with proper technique will usually have a smaller heat affected zone, and the discoloration will be even across the whole length of the weld. This tells you the welder started his weld, got it up to temperature, and welded at an even and consistent pace. So, you can deduce that the penetration of that weld will be pretty consistent all the way across. That is- if the weld is tied in correctly and has enough crown.

This takes some practice as it requires you to weld at a higher amperage than you may be comfortable with. This means you need to start the arc, form the puddle, start filling and get moving quickly, otherwise you can overheat the material. The other downside is if you are moving fast you may find it more difficult to stay steady. Practice makes perfect in this instance, and eventually it becomes natural. It can help to have an arm rest, or some sort of a prop to put your hand on while you're welding. Sometimes I use a clamp if I can make it work. I will just clamp a quick clamp or an F clamp to whatever I'm working on and rest my hand on it.

The key to consistency is comfort. If you are tense you are going to have trouble keeping a steady hand. So, do what you've got to do to keep your hand steady. Sometimes you might not have the space to clamp something, or you might not have something to anchor your hands on even. So, you'll have to do it freehand, or walk the cup if you are able. Whatever you've got to do to get yourself comfortable enough.

So, once you've figured out how you're going to anchor your body and hands, its time to look into grabbing your torch and making sure it's ready to go.

Now, I like to set my stickout based on the joint I'm welding. I'm little more particular if I am dealing with a fillet joint, because I like to drag the cup when I can for consistency. Normally I will take my cup and place it down into the joint until the edges rest on the two faces that are being welded. And I will adjust the tungsten so that with my cup touching the materials, the tungsten still has the arc gap that I am looking for. Usually then I will weld the gap the same way. Cup touching the materials, and just light up, and run the joint by dragging my cup across.

This means my edges will stay even, and my heat affected zone will be uniform, so long as the material is clean.

For a normal butt joint, I'd probably set my sickout to about ¼- 3/8". But sometimes I run it at ½". I've been known to go even more when I need it, so don't be afraid to experiment. You'll want to test it on a piece of scrap, and make sure that your gas coverage is still good. When you stick the tungsten out too far, it can keep the shielding gas from protecting the puddle, and you'll end up with a mess of porosity.

I like to keep my tungsten ground to a sharp point and I always try to keep the sharpening scratches in the tungsten running long ways down the tip. I sharpen them on a belt sander normally but have been known to use a bench grinder or a 4.5" wheel on an angle grinder. I like to get about 4 or 5 tungstens at a time and sharpen both ends and keep them around just in case I make a mistake.

You'll run through more tungsten, yes. But if you want perfect welds, you have to be willing to do everything perfectly. I'm super picky about how clean my tungstens are. I grind all the time. Some people see that as a sign of you not being skilled. I've worked in shops where people would watch you grind your tungstens, and if you had to make a second trip back, they assumed you couldn't keep your hand steady.

But a lot of times, the tungsten gets dirty just from welding. The puddle may have some dirt or oil that pops on you, and you pick up some crap on the tungsten. Maybe the heat rounded the tip out and the arc is acting funny, or maybe you did just dip the tungsten. It happens.

When you are just starting out, or are learning, you will without a doubt end up dipping your tungsten, or maybe dabbing your tungsten with some filler rod. So be careful with that, if you dip your tungsten a lot, maybe only sharpen one end of your tungsten so that you don't get it stuck in the collet when you get a glob on both ends.

Personally, I very rarely dip or get globs on my tungsten, so I sharpen both ends and keep a bunch of clean tungsten handy. If I find I'm having trouble with a wandering arc, or a hesitant arc start, I change the tungsten out, If I'm welding and I accidentally dip the tungsten or dab it with filler, I change the tungsten out. You will immediately notice what equates to about a 10-15amp drop in power when you contaminate the tungsten and continue to weld.

The filler material that was picked up adds resistance to electrode and it gives you a weaker arc. So, you will lose a lot of control over your puddle, and you won't have good penetration. Its best to stop and change your tungsten instead of running it. If it's a non-critical application, then whatever, its not the end of the world, but understand that you'll have less penetration where you welded with the dirty tungsten.

If you dip the tungsten and get it stuck, sometimes if you immediately start wiggling the torch you can work it out of the hole before it totally fuses, but once it fuses, you must break off the tungsten, and melt it all back into place, which in many cases is not allowed. So, you would have to grind and re weld that area.

So, do your best to not dip your tungsten. It can be tricky, because to get good arc control and quality, you need to have a very Tight arc gap – which is the distance between the work piece and the tip of your electrode, with enough practice you can get good with it. But when you're starting out, it can be tricky.

I like to keep my arc gap pretty much as Tight as possible always. You can sort of see what's going on by the color of the light coming off the weld. When you start getting too close, the light gets a little darker, and you will notice it in your hood.

I like to be just on that verge of it going darker. You can watch the arc pushing off your tungsten, and the arc force creates a dimple in the puddle. The tip of your tungsten is lower than the surface of the weld, the arc force pushes a dimple into the puddle that levels itself out when you pass the area. When the arc gap is this Tight sometimes you'll hear the material hissing and whistling a little when you weld. That's the sweet spot. If your gas is right, and your heat is right, if your sickout is consistent and Tight, you should be laying down gold coins on stainless steel.

As far as torch angle goes, most people recommend a 15-degree angle toward the direction of travel. I would recommend getting used to working at all types of crazy torch angles, both left handed and right handed, because there will be situations where you can't hold the torch perfect, either it won't fit, or your hand won't reach, whatever the issue is, there will be times where you need to break the rules, and you should be comfortable in being able to do so.

I can't tell you how many times I've saved myself enormous amounts of time on a job by being able to weld left handed. I might be laying on my belly, welding something at floor level right handed, but then have to make a weld in the opposite direction that would mean I have to completely reposition. Being able to weld left handed allowed me to not reposition and do the weld in much less time. So being versatile and able to adapt to the circumstances is going to make you a much better welder.

For the most part, you will want to keep your torch angle consistent, and angled in the direction of travel, however there will be instances where you might not be able to feed the wire due to an obstruction, so you may actually have to weld backwards. This sounds easy, but its not always easy. You need to keep gas coverage, so if you're doing things backwards, its easy to lose your gas, and get porosity.

Eventually when you establish good puddle control, you get a feel for it and can weld in pretty much any direction without getting pinholes or porosity.

There are several techniques for working the puddle and adding filler. Each of them is going to give you a different look and each has its place. The most common methods I will use on a day to day basis are the dabbing method, which is pretty basic, and is probably what 95% of people who weld will use for most their projects.

Basically, you line up your tungsten, start the arc, give it a second to allow the puddle to form, and then start dabbing and moving forward at a consistent speed. This will produce beautiful welds with great color if you have your heat, sickout, and gas set properly. It will produce a consistent bead with consistent penetration and works great on fillet and butt joints on pretty much any material.

The downside to this method is you are not going to get as deep of penetration as you possibly could, and since you are moving at a consistent rate of speed, the puddle will tend to take only as much filler as it can, so you may have low weld deposition rates, and may need to run a few passes using this technique depending on how thick your base material is, and what the job calls for.

For most lighter duty welding, this technique works fine for single pass welds. This technique requires you to be pretty stable in terms of positioning, as you are going to need to move your arm in one single solid motion, and anything that Might stop you from doing so will mess up your bead.

You can either fab up an arm rest to clamp onto your table, or you can sometimes get away with clamping a clamp right to the work piece depending on the situation. This will help you with balancing your hand and can give you a nice surface to slide across depending on how you construct your arm rest.

It doesn't take long after you start the arc to get the puddle moving. If you are using enough heat, it's almost immediate. On say a fillet weld also called a T joint, I like to adjust my stick out on my tungsten to be just ever so slightly higher than where I want my weld to be. Using this technique, with say a number 8 cup, you can set your sickout and just ride the edges of the cup on the material to keep you steady, and just drag the cup across the joint in a single motion as you dab your filler. This will make for almost robotic looking welds. You need to be consistent in your travel speed and your dab timing, and because you're riding the cup on the edge, your arc gap and weld size will be consistent.

This is also a great method to use if you are using a welding positioner. If you're not familiar with welding positioners, they are also called rotators, welding lathes, etc. Typically, they are motorized, and run off a foot pedal. You chuck your piece up, start the arc, let the puddle form, and hit the pedal, and the piece begins spinning. It moves consistently, and all you have to do for the most part is just dab, dab, dab, your filler at a consistent rate.

But beyond the dabbing method, there is an approach I use a little more often that comes in handy when I'm welding a thicker material and maybe I have a gap I need to fill. And what I basically do is sort of a technique I borrowed from MIG welding.

I will start my puddle, and add some filler and then step forward, add another dab of filler then step back about half a step, then forward, Dab, Back, And I kind of work the puddle around in a circle, and forward and backward with the torch.

Basically, moving forward, drawing in some rod into the puddle, then pushing the puddle back into a neat stack of dimes.

This technique produces very nice looking welds, but they have more of an organic look to them, the edges won't always be completely even, But the advantage to using this method is that you can fill gaps rather easily because you are keeping the puddle moving as opposed to keeping your heat in one stationary spot. It means the filler material wont flow as freely, which allows you to build up weld size and fill gaps rather quickly. It's even easier when you use larger diameter rods. The extra size will help fill more of the gap per pass.

This method usually uses a dabbing technique on the filler rod. If you try just laying the wire into the joint and running over it, what happens is a lot of times the wire will weld itself to the base material when you pull the puddle backward to make your ripple, so there's a little technique involved.

It can be a little bit annoying to have the rod welding itself to the puddle every two seconds, so you need to be careful about the angle at which you feed your rod, and you need to pull the rod back up out of the puddle before you start to push the puddle backwards.

Whenever you do this, it's important to keep the end of the hot filler rod within the gas shielding, otherwise the end of the filler will oxidize, and you may have trouble with porosity or you may get some trash in your weld when you start welding again. So, it's always good practice to keep everything shielded until the weld is has cooled.

This is one of the biggest reasons people's welds don't look great. If you are welding, and this goes for any material pretty much, and you remove the filler rod from the shielding gas before its cool (when the end is still red hot) that red hot ball of metal will self-contaminate, and cause issues with cleanliness in your welds. Especially with aluminum. Aluminum can be a pain in the ass when it's not kept clean.

Also, this back forth technique, I wouldn't recommend it on aluminum only because you're moving around a lot, and it will reduce the penetration into the base material and potentially cause porosity if you aren't careful.

This can be true for all materials. If you move the puddle around too quickly, it can solidify before you boil out any air pockets, and you end up with porosity.

The third method that is most commonly used is called the lay-wire method. It's a pretty common pipe welding technique and is great for speeding up your welding, while still getting a weld that has great penetration, and has a nice overall appearance.

Basically, you have a joint, and most likely it's going to have some sort of a bevel on it, (but it really doesn't need one if your technique is good).

The idea is that basically you would lay the wire right into the root of the bevel and run over the joint with the torch in a weaving motion, tying the two pieces together. The heat and travel speed will automatically draw in the appropriate amount of wire.

You can feed the wire as you normally would with your hand, and sort of push the wire into the puddle, but this isn't always necessary. If you are using the right size wire for the bevel you are working with, you can pretty much stick the wire into the joint and put a little downward pressure on it and as it melts it will walk itself up the bevel and practically feed itself.

You will basically start the puddle, give it a couple dabs to get the puddle to the correct size, and then stick the filler in, and work the whole puddle back and for the between the two pieces, your travel speed will determine how much of the weld is filled. The slower you move, the more rod the puddle will draw in, the bigger the weld. The faster you go, the less material is drawn in, the lower the weld.

When I do this method, I like to put a little twisting rotating force on the filler rod, to keep it sort of twisting against the bevel. If you don't keep the rod moving, it can fuse itself to the joint. So, with my thumb of my hand that I feed the wire with, I just apply a little rotational pressure, and it keeps the rod free of sticking to the puddle for the most part.

I use this method most if I am welding a gap, whether it's an intentional gap, or if there is a gap due to a mistake in cutting. Because you are moving the puddle back and forth across the joint, it dissipates the heat better. When you are filling large gaps, sometimes the biggest issue is that your heat is too concentrated, and you begin to overheat the material. And if you take it too far, you end up with a big bright yellow ball of molten steel, and you add that last dab of filler and it just drops out on you.

So, I like to work the puddle around, sometimes backing off the pedal if I have one to reduce the temperature and give it a second to cool, and if I don't have a pedal, I will long arc it a bit which will also reduce the heat input.

What I mean by that is I will actually pull the tungsten further away from the puddle mid weld. I usually also like to move it away from the center of the puddle when I do this. the idea is to just get the focus of your heat away from the hottest part of the puddle.

The lay wire method works great if you get the temperature correct and can be quite a bit faster than the dabbing method. You will commonly see this method used when welding pipe, along with a technique called walking the cup, which I touched briefly on earlier.

You can also do the back and forth motion with the torch if you are using small wire and are travelling fast enough to not get the filler rod stuck to the base material. A little bit of practice will make that pretty easy.

laywire will produce very nice looking flat, consistent welds. They will not usually have that stack of dimes look that many people are looking for, but the advantage is you are keeping the heat focused, and consistent by manipulating the puddle between the two pieces. It keeps the heat focused and gives great penetration. I would recommend this method if you were doing say a multipass weld on some thick wall tubing or pipe, and you wanted to get the weld in quickly.

If you weld pipe too slowly, you can overheat the material, which again weakens the pipe. That's a big no-no. Also, it speeds things up, because you need to allow the pipe time to cool between passes. If you run your root pass, then immediately jump in to run your fill pass without cooling down the joint, you'll end up cooking the material. The puddle will flow much faster than the first pass did, which means you'll have uneven weld deposits, so part of the weld might fill up most of the bevel, but then the hotter it gets, the less the metal fills the bevel, and the more it wets out.

Lay wire keeps heat input to a minimum by increasing travel speed and overall decreasing the time that you spend putting amperage into the material. This allows it to cool faster between passes, making a much more solid weld with less of a chance of overheating the material.

Aluminum Technique

Aluminum is a tricky material to weld. It reacts much differently to heat than other materials. You will notice immediately that it does not turn red like steel or stainless steel do when it gets hot, instead it gets really shiny.

The problem most people have with aluminum is they weld too slow. As I discussed before, you want to minimize heat input on your parts, with aluminum if you weld slowly, that aluminum is absorbing all that heat, and it is dissipating it very well. So, people will notice that even after just a couple of tacks, the entire part will be hot.

So, the issue a lot of people experience, is the aluminum will get too hot, and the whole part pretty much will just melt, or a large portion of it. And it happens quickly. One second you are waiting on the puddle to melt, the next second you are starting from scratch because half the part melted away.

This is typically caused by one of a couple of problems. The first and most obvious being the machine is either too weak, or not set high enough. If you for example need 180 amps to weld a piece, and your welder can only do 160, you'll sit there for a long time holding the arc on the weld joint waiting for it to form a puddle. But since the machine doesn't have the juice, it will eventually overheat the part and melt the whole thing.

The second issue that people face is improper prep of the material. The oxide layer on aluminum melts at around 5,000 degrees, while the actual material melts at around 1200. So, if you don't properly clean the oxide off the aluminum, you will be doing the same thing, sitting there until the cows come home waiting on a puddle to form. meanwhile the oxide is nowhere close to melting, but the material below it is just about there. And in a split second, you see the surface wrinkle, and its all over.

With aluminum, I typically use a pretty methodical system. It works for me, so it's what I do. First things first. When you weld aluminum, you are fighting a layer of oxides that melts at many times the temperature that aluminum itself melts at. So, you need to remove the oxide layer first. You can sand this layer off with a DA sander, or a piece of sand paper. You can use a wire brush and some scotch bride, that has been my preferred method lately. Run the whole joint with a wire brush, then go over it with a scotch bride, and wipe clean with lacquer thinner or acetone.

Make sure your pieces are as Tightly fit as possible. While it is possible to fill gaps with aluminum, I don't recommend it unless you have absolutely no choice. If you can spend a few minutes getting your fit up perfect, you are better off. It will weld much nicer.

If you are welding thin aluminum, make sure to tack your parts together often before welding. Aluminum does shrink when you weld it, and not tacking it completely can cause some unfavorable distortion. You'll find gaps opening where there weren't any, and all sorts of bad things. So, make sure you spend some time and put a good amount of tacks around your workpiece to make sure everything stays right where you want it. This is especially true with butt joints or outside corner joints.

Say you are welding a couple pieces of .040" aluminum. It's a butt weld, it's about 12" long, and you've tacked it at both ends with a Tight fit. If I start my bead at one end, the two pieces can move freely between the two tacks as I weld, you will find that about half way through the weld, the pieces start to separate because the heat has caused them to bow out and open a gap. To avoid situations like this, don't be afraid to put a tack every couple of inches or even less.

Just be mindful of the fact that you will be welding the piece, and if you get too crazy with adding filler and making big blotchy tacks, it's going to be hard to make an even consistent bead and blend the tacks in smoothly. So, for this reason you should keep your tacks as small as you can, making them only as big as needed to hold the pieces together, and when you go to run that final pass over everything, spend an extra second on your tacks and try to re-melt them so that everything becomes one solid consistent bead.

For my personal preference with aluminum, lately I have been running 2% lanthanated tungsten ground to a point with aluminum. Some people will recommend grinding a flat spot on the tip, you can do this, I don't, I find that the arc will naturally round out the tip of the tungsten.

If you are running an older transformer machine, you will want to try pure tungsten, and ball the end by blasting the tungsten with DCEP on a piece of copper. They tend to run better with that setup. Newer inverter-based welding machines, however, can run just about anything on all processes. Some people prefer lanthanated, some people prefer e3, I like my thoriated. They're cheap and usually pretty readily available, plus they are tried and true for many many years.

So, let's say we are going to weld some aluminum plate. ¼" thick plate to be exact with a Tight-fitting butt joint. Let's say it's a critical joint, and I want to make pretty good penetration. My setup would be pretty basic; AC, about 200 amps, and I wouldn't really mess with ac balance too much because we are dealing with clean material and I've prepped it well, so the cleaning action isn't really necessary.

I would wire brush the entire joint to remove as much of the oxide as possible, then scotch-brite over that to make sure. Next, I'd wipe all my joints with lacquer thinner, I would be using a number 8 cup with 20-25cfh, I sometimes like to use a little more gas with aluminum. Tungsten Stickout would be around 3/8-1/2". Enough to clearly see what I'm doing.

Light the arc up and work it back and forth from piece to piece. You will see the edge of the pieces start to get a little shiny. Keep working this spot until it starts to bubble up a little bit. When the edge is melted over and looks very shiny, that's when you know the puddle is ready to dip. Dip your filler, get your puddle to the size you want it, then start moving forward, dabbing consistently and evenly, travelling consistently and evenly. If your weld starts to get wider and wider, it means you are putting in too much heat or moving too slowly. Next time spend a little more time on your puddle, so it wets out fully before you start travelling.

When you get your heat dialed, you should be able to lay consistent welds without having to vary your pedal too much. In any situation, regardless of material, if you find your machine is running too hot, try coming down 10 amps, if that is too much, go back up 5. Slowly play around with it and adjust in 5 or 10 amp increments until you get it set close to where you want it. With aluminum, the weld should look silver and almost polished. If its gray and has specks and black crud on it, it's not a good weld.

If you get a bad weld with aluminum, start over. Grind it out and try again, because aluminum is pretty prone to cracking. Especially if the metal is contaminated in any way. I see this a lot on alloy wheels.

I often get people asking to weld alloy wheels. And when they bring them in, a lot of times they've already been welded on a few times. They pretty much always re-crack right down the center of the weld.

So far of all the wheels I've welded, I have yet to see a single aluminum wheel that was actually done correctly. Most of them have boogered up welds that they didn't bother cleaning or slicing out the crack. Usually the crack is loaded up with all kinds of dirt and trash, and the most common failures I see are easy to avoid if you take a little time to do things the right way.

When you take your time, and everything comes out nice, they last. I have not had a single wheel come back broken. However, if you don't do it right, and you allow contamination into the weld, it will almost definitely crack again once you hit a pothole or whatever, and they usually don't last very long. I did an extensive write up on welding cracked aluminum wheels on my website.

Pulse Welding

Here's a pretty misunderstood topic when it comes to TIG welding. Pulse welding. A lot of people have absolutely no idea how to use a pulser on a TIG machine, and that's unfortunate because you can really do some neat things with a pulser.

As we discussed before, the main job of the pulsing circuit is to basically to pulse the current, as if you were pumping the foot pedal the pulse circuit on a welding machine this automatically, and to whatever parameters you set it for. You can adjust the high and the low end of the pulse, as well as the pulse frequency.

Typically, you will have 3 controls to operate a pulser. There will be a frequency control, could be labeled in HZ or could be labeled as PPS (pulses per second). This controls the overall speed of the pulse. The other two controls will set the limits of the current. One will set the low end, and one will set the high end, so you can set it to pulse for example between 30-70 amps, or 20-80 amps, or 50-70 amps, etc.

You are setting the high limit and the low limit. The lower you set your low-end limiter, the larger a sweep there is between currents, which means less heat input, but also a colder puddle, so in some cases you may have to turn your machine up quite a bit to compensate for the sweep of the pulser and the reduced heat input.

Older transformer welders like the miller Syncrowave 350dx will have a built in pulser, but it has a very limited pulse frequency. Many people find it difficult to use, because around 1pps you can get a very nice rhythm going for dabbing your filler, and it makes it easy to make a nice stack of dimes, but crank it up a little more than that, and you can't find that rhythm, and it makes it difficult to predict when to add your filler. These machines, the pulser is pretty much just used for that purpose, for laying out nice dime stack looking welds. I'm sure there are uses for some of the higher pps settings on the transformer machines, but I haven't found a need for it.

On inverter machines, it gets interesting because you typically can go up to 200 and sometimes more pulses per second. This comes in very handy for keeping heat down when you want to avoid warping or discoloration, while still getting decent penetration.

An example where pulse saved the day for me, was when I was doing some stainless-steel counter tops a couple of years ago.

The architect wanted a mill finish on the top. That means no sanding, no scotch-brite, no grinding, nothing. They wanted the original factory stainless steel mill finish. The counter top was made from ¼" thick 308 stainless steel, and was laser cut for a sink I believe and some lights.

The edge of the counter top was a piece of ½"x3" stainless steel flat bar welded to the underside of the ¼" thick stainless-steel top with the weld ground smooth, so it appeared to be a 3" thick slab of stainless steel. The flat bar for the edge had to be welded to the ¼" and it had to be square to the counter top. It needed to be penetrated because the weld seam was going to be ground flush, and we didn't want cracks to develop after grinding.

To keep the bar square to the surface of the counter top, the idea was to weld on the inside and the outside with a series of stitches and tacks placed strategically to help pull the bar in one direction or another.

The two big issues with this project were A) we had to keep the mill finish, meaning no discoloration from welding, and the material was only ¼" thick, so it really doesn't take much to discolor it, and B) the counter top was about 10ft long, and we had to make a weld the entire length of it, on one side. So, warping was a big concern too.

So, what I ended up doing was after getting the edge bars tacked the whole length and squared up, so it was exactly where I wanted it, I was using a miller dynasty 350, and I set my machine to 190 amps, 200PPS, and I don't recall the settings for the high and low limiter.

I put some back spring in the piece by putting a spacer in the center, and then clamping the ends across a straight edge. I believe I used like a big 6" piece of square tubing we had in the shop for a stair stringer on another job. The idea was to counteract the warping that would happen from welding it. I do this pretty frequently with stainless steel to keep parts coming out straight.

I will kind of guess how much and where it's going to pull, and counteract that by preloading it in the opposite direction it will pull in.

I also had a long 1 ¼" solid brass bar that I clamped on the surface of the counter top, right above where I was going to weld. This was to draw heat out of the ¼" material to avoid discoloration.

I drew a tape and marked my weld seam out in 4" sections and skipped around using the dynasty with the pulser at 200ppm and 190 amps. My welds came out with a strange puddle shape due to the pulser, but the welds were gold, and said and done, between 2 counter tops that I did there was no discoloration on the surface of the counter top, and over 10ft it only warped less than ¼", which was nothing considering it was getting glued and clamped to a wooden structure. At the time my boss said they were the best counter tops he had seen as far as warping goes.

For most people, a situation like this will probably never come up. But if you tried welding it with straight polarity dc, it would have warped much, much worse, and there would have been discoloration on the surface everywhere there was a weld, which would have pretty much killed the look the architect was going for. It's good to know these types of things can be done because who knows, you may find yourself one day in a similar situation where this will help you.

Feel free to experiment with things like using brass and aluminum to draw heat out of the workpiece to reduce warping. I have used this technique with success countless times. Copper and brass are your best choices for heat sinking, but aluminum works as well.

I made some brackets once that were a ½" thick piece of stainless about 2" wide by 5" long, and there was just a little ear welded across the 2" face right in the middle I took a couple of small strips of copper sheet metal and placed them right beneath the weld, and then on top of a ½" thick copper bar. I clamped everything down with C clamps, and the sheet metal strips acted as spacers to back spring the part. Welded no pulse straight dc, but the copper pulled the heat right out of the stainless. No warping at all. This is especially useful on stainless steel, as stainless tends to hold onto heat much better than steel, so the copper draws the heat right out of the stainless and really cuts down on warping.

You should try to employ these types of techniques whenever you work with stainless steel, as it can be a tricky metal to work with if you don't know what to expect. However, these techniques can be used on all materials where you are concerned with warping. And with a little bit of practice, it becomes second nature.

There have been several instances where I was faced with keeping warping to a minimum and using the heat sync method, I was able to keep the parts as straight as an arrow.

Weld Sequence

A topic that many people don't really touch on is weld sequence. This is one of those things that comes with experience, but for most people, you end up having to screw up a lot of stuff to get the experience needed to have the forethought necessary to consider weld sequence. There are times when you can spend all the time in the world prepping and getting comfortable, and your welds look beautiful, but the part ends up in the scrap bin because you didn't plan your weld sequence properly, and now the part is warped, and twisted every which way.

So, let's take that square frame we were working on earlier in the book. 30x30 stainless steel square tubing with 45* miter cuts at each corner. My parts are clamped to a table using a straight edge and a square to make sure everything is where it should be, but my corners still have a slight gap at the inside corner of the tubing. The outer corner is mint, everything touches nicely, but my gap is about 1/32" on the inside corner, so my cuts are slightly off from the 45* mark.

So, my options are to go ahead and sand the miters to the correct angle, however you would lose some material making it a little less than 30”. If you need to keep a consistent inner dimension, as is often the case when building frames, you pretty much are stuck with what you’ve got.

So, in order to weld this frame together, keeping the corners square and not warping it, you will need to plan your weld sequence. In earlier chapters, we talked about how you could lock in the gap on a corner joint by placing tack welds along the gap. So, let’s assume we have already done this, and the frame we built measures 30x30, its currently square and all my outside corners are fit-up Tightly. My inside corners have a 1/32 gap at each of them. And I have tacked all 4 corners of each miter.

So, assuming we are perfectly square I will usually just pick a corner and start welding. I will start my weld at the outside corner and run my bead inwards toward the inside corner. I would keep the puddle consistently moving and do the weld in one pass. Now if you check square you’re going to notice the part pulled together a bit, and it’s no longer square. Now some people will spend lots of time setting up jigs and clamps to try to reduce movement, but when it comes to stainless steel, it's going to pull no matter what. That's just how it is. So, the way to reduce distortion to a minimum is to use it to your advantage.

So now that we have one corner welded and we know it's out of square, we must figure out how to correct the issue before it becomes permanent. So, you'll want to do the exact opposite of what you just did. You want to counteract that pull by welding the opposite corner now in the same inward direction.

Keep in mind, your weld direction is very important. If your first weld was made from the outside corner to the inside corner, if you start your second weld on the opposite corner of the frame at the inside corner of the joint, it's basically going to want to pull opposite, which will pretty much lock your problem in permanently.

So, you want to start at the outside corner and weld inward, then do the opposite corner, then flip it around and do another corner and repeat. Until you've done all the outside faces. Then I would weld up the outside corners, because I know these are going to have less of a pulling effect on the frame than the inside corners will. I will finally weld the inside corners last. They are going to have the most "pull" when you weld them due to the angle of the cut, so you save those for last. You want to make sure your using consistent heat, you want to make sure your welding in the proper direction, and assuming you did everything right, the piece should come out perfectly square without using clamps to hold it.

You will get used to the way the metal moves in time. It becomes very easily predictable and this all eventually becomes second nature. But this is a very important principal in welding and metalworking that many inexperienced welders do not account for. When you weld something, or even just tack it, the part is going to pull towards the weld side. This can work for you, or it can work against you. Which way that works out depends on how far ahead you think. So, planning your weld sequences, and coming up with a plan of attack before you turn the machine on is good practice to get into.

Let's say you have a post, say for a railing. it's a piece of 1 ¼" square tubing about 34.5" long, and you are welding it vertically to a piece of 4" square tubing that lays horizontally. Your post was cut on a standard miter saw, and the cut is not quite a perfect 90. You really need this post to be plum otherwise it's going to look out of place, so in this instance, you would want to know exactly where to start welding and which direction to weld in to counter act the pulling effect against the imperfect cut.

So, you plum the post out, so that It's exactly 90 degrees perpendicular to the 4" rail its being welded to. Only 1 corner of your post is touching the surface of the 4" tubing due to the cut being a little sketchy. You can go ahead and grind that post and make it perfect, but you may lose a little length. So, let's say for the sake of argument you don't want to lose that length, and you need to make it work. I have in the field many times had this situation where, if I lose 1 16" of an inch grinding to straighten a piece, it's going to screw something else up, so you need to be a bit methodical about how you weld the part, or it will go in crooked.

With railings, even though 1/16th of an inch doesn't seem like much, if your post is short, the top bar needs to be pushed down to meet the post when you weld. So, when you see it from a distance, it will look all wavy, especially if the post is towards the middle of the rail. So, most of the time when I built railings, I would try to fill the gap and keep the post square and plumb.

In this instance, I would start by getting the post plumb using a level on 2 faces. I would tack the one corner that's touching, and put an extra dab of filler on it, just because you may be moving it around a little.

So, once I have my first tack made, ill re-check level, and it should have pulled out a little bit towards your tack. So, you'll want to pull against the tack to bring it back to plumb and get another tack on it. Usually I'll go for the opposite corner if I can. Then re-check with a level, tack the next closest corner, re check with a level, make any necessary adjustments and tack the final corner. You should still have gaps at the bottom of the post, and you most likely will have had to put in a few dabs of filler to make it work. But the post is now tacked plumb and at the right length.

So now say you want to fully weld the post. Let's assume it's perfectly plumb as you tacked it, but still has a small gap on 3 of the 4 corners. If you are not careful about how you weld this, you can end up pulling the post towards your gap, and leaning it in quite a bit, so this is where you need to be careful.

You need to pay attention to your weld direction and you want to weld in a direction that will lean the post away from the gap, so you would start your weld at the corner with the biggest gap.

Keep in mind here, I am talking about small gaps, you don't want to go much bigger than the thickness of the material, because then it gets hard to fill. So, assuming our post is tacked right where we want it, and the gap is say 1/16 at its worst point, I can easily fill a gap that size with a 1/16 rod, So I start my weld by placing a couple of tacks on the flat faces of the post to lock in that corner, so that when I start my weld it can't pull down much more, which would pull the post out of plumb. I would start my weld as I said on the corner with the largest gap, and weld towards the corner with the smallest gap, stopping to check plumb between each weld.

You pretty much choose your sequence based on what the post is doing, you can normally predict it ahead of time, but it's nice to check your work along the way in case something screwy happens, and sometimes it does!

So, you might find in this instance, the weld sequence would be to start your weld at the tack with the largest gap, and weld one face towards the opposite corner. Then coming back to the same starting point weld the other face in the same direction. You would now have half of the tube welded. The other half of the tube might be best started from the opposite corner (the corner with the smallest gap) and welding away from that corner towards the center of the post. What this will effectively have done, is locked in the corner at the appropriate gap to keep the post plumb, and then by welding in specific directions you counter act the pulling effect and use it to keep the post pretty plumb.

This may sound a little confusing, but with time it will become pretty obvious. But this is the type of knowledge that separates rookies from real fabricators. Anyone can just take a bunch of parts and slap them together and weld them with no specific tolerances. But being able to work within less than $1/16^{th}$ of an inch in most cases will make your work look much better.

And in my career, truthfully, most of the work I've done didn't need to be that accurate. A lot of the prints would have like 1/8" tolerance or sometimes even more. But why shoot for "close enough" when you can take a little time and do it right? Eventually it ends up not taking any extra time, because you already know what to expect.

The Effects of Heat

With stainless steel, you are always going to be fighting heat distortion, and this same technique can be applied to other materials as well, it's just more pronounced with stainless steel. One thing that's nice about using the TIG process on any material is your ability to heat up a weld without adding any filler. So, in an event you tack a piece in place, and it contracts more than you were expecting, you can always apply a pressure in the direction you need the post to move, and heat up the tack, which allows the post to move a little.

You can do this alternating between tacks to move the post right where you need it, then lock it down with a couple more tacks and then fully weld. These extra steps would not be necessary if your cuts were perfect to begin with, but in the real world this isn't always the case, so it's best to be prepared to deal with unfavorable conditions.

You can save yourself a lot of time on cutting apart joints and tweaking and re-welding posts that pulled out of plumb if you just take a little time to think ahead and plan your weld sequence and direction. I used this technique extensively when I was doing railings, and architectural work.

A lot of times we would be installing posts in the field and welding them to a metal stair stringer, and you really don't have anything to go off, you can't use a square to check for plumb, you pretty much need to use a post level, and plan your welds to lean the post right where you need it to be. Sometimes we would purposely tack the posts slightly off plumb, and then weld in the direction it needed to tilt, and it would pull itself straight. You eventually get a feel for how much the material will move, and it becomes easy.

Now let's say we took the same post we talked about earlier, welded to the same 4" piece of square tubing. Now let's say with the cut being off the same amount, we tacked the post in the same way, but we welded everything starting from the corner that was touching, working towards the corer with the largest gap.

What would happen is what I like to call the "zipper effect". I don't know if there is an actual technical term for it, but what happens is, if you are welding in the direction the gap increases, it's going to allow the post to pull incrementally with every dab of filler.

So, what happens is you end up pulling the post over several degrees. And while yes, the opposite corner (with the large gap) was tacked and locked down, Welding with the direction of the gap causes it to incrementally get Tighter with every dab of filler, so when you come down to that last dab where you melt the original tack, it will squeeze Tight on itself once the tack is melted. it may not completely pull down to the base material, but you will almost always have a bit of lean going on. At the end of the day, 1 1/6" can really make a big difference in how your work comes out, so it's important to be as precise with your fit-up and welding as possible.

You can use these principals to correct things that are welded fully and are slightly out of adjustment as well. With stainless steel, whenever you weld, there will be some degree of contraction. The degree of contraction is based upon your technique and your machine temperature. So, the key here is being consistent. Therefore, I like to recommend using a finger control over a pedal.

Let's say I have my machine set to 70 amps. I am welding some 1/8" thick 308 stainless steel tubing. I know I can predict how much it's going to pull if I can be consistent with my heat.

So, if I'm using a finger control, which mine happens to be a switch, I know that assuming good technique and a consistent arc gap, I am going to have consistent penetration throughout the whole weld, because there is no variance in the current. It's just on at 70 amps or off, so we can assume that the entire structure, if welded at 70 amps should be penetrated equally, if technique is unchanged.

Now let's say I'm welding in some posts for a railing. And I'm starting from the bottom and working my way up. I weld the post onto the stringer and notice that it's still a little off to one direction. Now you can go ahead and whack it with a mallet and try to move it a little, but if that's not an option, or if it's not enough, you can set your machine a little bit above 70 amps, and what you will end up with is a hotter weld, and deeper penetration, which translates to more pulling force. So simply re-welding a joint with a little more power can correct pulling.

So, say a post is leaning to the right 1/8". If you set your machine to say 75 amps, and start welding towards the direction that the post needs to lean in, in theory you will be penetrating deeper into the root, which causes more pulling on the material, and since you are welding away from the leaning direction, the "zipper effect" will cause the post to slightly move in the direction of travel little by little as the weld puddle freezes. Sometimes you need to do it a couple times, and sometimes you have to pull or push on it in the direction you need it to move.

Sometimes all you need to do is weld in the proper direction. sometimes you must pull on the post a bit while you're welding, but you can usually get parts to move one way or another at least a little bit even after they are welded just by re-welding them in the proper direction with a higher current than what the original welds were done at. This is one of the best things about TIG welding, you can still adjust things relatively easily by heating and separating tacks as opposed to grinding and cutting tacks like you would need to do with a MIG or stick welder.

A lot of times if I tack something in and notice its wrong. I'll just heat the tacks and pry the joint apart, then I can come in very lightly with a flap wheel, shave down the tacks to where they aren't in my way. Re position my post and tack again. It makes your work look much cleaner than if you use a grinder to cut the tacks apart and leave cut marks on the material. There's nothing more disappointing than seeing a project come out perfectly, but there's just that one spot you hit with the cut off wheel and left a gouge.

Like I said earlier, being a good welder doesn't always mean you get it right every single time, it means you're smart enough to cover your tracks, and make it look like you never screwed up in the first place.

Mistakes happen, were all human, we have good days and bad days. Sometimes I didn't get to sleep until 3am so I would wake up for work the next day and there just isn't enough coffee on earth to keep me going without making SOME kind of errors. As long as you're not wasting all day fixing your mistakes, and you can do it in a way that doesn't affect the outcome, you're doing alright.

Work Habits

When it comes to handling the torch, there is no right or wrong way to hold your torch. So, get comfortable holding the torch in as many different hand positions as you can think of. They all come in handy in the field when you must make a weld in a weird position.

Having a good grip and control of your torch will help you keep control over how the metal is laying down in the joint. You don't want to be holding onto the torch with a death grip. you want it to be pretty loose and maneuverable in your hand, so you'll want to experiment with slinging the cord over your shoulder or arm to reduce pulling on your wrist. This trick is especially handy if you are walking the cup. It can make the torch feel as if it's part of your arm. You don't feel any pulling or drag from the cord, and the result is a much more free and movable wrist.

Taper on and off. When you first start your arc, if possible, (if you are using a pedal) its good practice to start your arc low, and slowly and gradually increase the amperage until you are up to welding temperature.

The reason for doing this is to give the puddle a second to boil out any crap that might be inside it.

If you are starting a weld on a piece of material, that may or may not have been cleaned properly, there's a chance there could be some crud on the material, OR in the event that you are welding carbon steel with mill scale, as I talked about in the beginning of the book. If you don't taper the current in, sometimes that crud or mill scale will pop, and cause the puddle to blow back onto your tungsten. This contaminates your tungsten and you pretty much need to stop and clean the weld again and start over. It's a pain in the ass.

So, if I'm welding over mill scale, I will taper on, slowly ramp up to welding temp, (which I will normally use quite a bit of extra heat if there is mill scale) and basically weld slowly watching the puddle. You will see all kinds of trash floating and spinning in the puddle, and you're waiting for it to die down as much as it will. You won't get it to stop floating around but give the puddle a couple seconds to stabilize and slowly work forward adding rod, and keeping the heat consistent, to boil the crud out of the material.

When you have reached the end of your weld, slowly circle around, add an extra dab or two of filler and slowly taper off the current. The reason for tapering off at the end of a weld is to avoid cooling the weld too quickly. As we discussed earlier, when you heat metal it expands, and when your cool metal it contracts. Well when a weld cools too quickly, it will form a pinhole which may have cracks on the inside which can't be seen by the eye.

Pinholes and craters are bad news, and you want to avoid them at all costs. Any welding job that has a QC inspector or any sort of CWI on site will not pass anything with pinholes or craters, so pay attention to these, as people will be looking for them.

It is also important to watch out for undercut. Undercut is a gouging that happens at the edges of the weld. Sometimes people refer to it as underfill, however it's not quite the same. Undercut is caused by a few things. One being too much heat, the metal flows too fluidly, and doesn't tie into the base material properly. Another reason is mill scale. If you are welding with mill scale, you most likely will get undercut around your welds. This is because the filler cannot adhere to the mill scale, so it needs to cut deeper beneath the scale to fuse properly. And you are left with a ridge at the edge of where the puddle met the mill scale. The way to prevent this, is to clean your welds properly and remove all mill scale.

In an event where you are forced to weld over mill scale, Which I have had many situations where I was forced to do as an architectural welder, you will be forced to live with a small amount of undercut, as you cannot get a proper tie in without grinding the scale off.

Underfilling can also be a cause for undercut, but typically it's when combined with too much heat. Too much heat, not enough filler = undercut. Proper heat, not enough filler = underfill. Either way, whichever you want to call it, It's the same idea. There is a low spot at the edge of the weld, and there should not be.

A CWI will come in with a flashlight and shine a light on the weld. If he sees a shadow, he's going to fail the weld for undercut, weather you want to call it underfill or undercut makes no difference. (I have seen this argument play out several times on the internet, and it's ridiculous.

a failed weld is a failed weld no matter how you want to call it).

So, to avoid undercut all together, it's important to follow good prep procedures. Make sure no rust, oils, dirt, grease, debris, paint, adhesive etc. are on the materials. Clean everything properly. Prepare your gaps properly. Use proper technique, and make sure you are using a large enough filler rod. If the rod is melting away before you get it to the puddle, you are either using too small of a wire, or your torch angle is too steep. Or too much heat, Or a combination thereof.

Pay attention to your edges of your puddle. Watch the puddle tie into the base material on the top and the bottom of the weld, make sure its tied into the root properly. A lot of times undercut is caused by travel speed, and if you slow it down just a hair and let the puddle suck in a little bit more rod, you'll see that undercut disappear.

In addition to keeping your parts clean, it's a good idea to wipe down your filler rods before you start using them. Wipe them with acetone or lacquer thinner to remove any excess grease and oils. You will see how dirty they are when you look at your rag. It will make a big difference.

Do this with all materials, especially aluminum, because aluminum loves to react to trash in the puddle, so you want to introduce as little crap as possible to the weld puddle. You may also want to go as far as keeping a clean set of gloves just for welding and use another set for all your fitting and fabrication. Keep all your rods stored in a clean dry place so they don't start to corrode, and always use clean tungsten, never waste your time trying to weld with a cruddy tungsten. The result, even if it looks okay, will be a substandard weld that will not have as deep of penetration as the rest of the welds which were done with a clean electrode.

The most important part of TIG welding is to experiment and try your own ideas. The only wrong ideas are the ones that don't work. And every welder has his or her own views and outlooks on handling a given scenario.

There really are no right or wrong ways to get to the desired result. However, as a welder, if you are going to be working for an employer and not on your own, you will find that the biggest concerns your employer has are speed, speed, and speed. They want everything done in half the time it should take, so that you make them more money on the job. And there's nothing wrong with that, if it pushes you to figure out how to creatively make things happen without cutting corners and still save time. You will end up doing alright. Just don't take it to heart and start cutting corners and producing substandard work. Always take your time. Always do the most you can to save time on fabrication and welding.

But always keep in mind, at the end of the day. if a company gives you a job, and you are allowed 10 hours to complete it, and you do it in 9 and the customer loves it, they still make money, and they Might bitch at you to speed up, but ultimately, they can't say anything. But if you cut corners and start slacking on your welds and your fit-up, the overall result is going to suffer, and it doesn't matter if you did the job in 3 hours, the company is going to be pissed when they have to explain to the customer why their part came out like crap, and how quickly they can have it fixed. You'll have a hard time keeping a job at a company if you can't produce consistent quality work.

And so, while everyone wants to work on laying out dime stack welds, I say you should focus on making even uniform and consistent beads with proper tie in, proper starts and stops and no craters or pinholes or undercut. Even if your welds don't have the stack of dimes look, if you follow those guidelines you will have good looking welds and people will compliment them. And in time, as your technique progresses, and you get better at feeding wire and keeping a rhythm for your dabbing, you will start to produce more and more perfect dime stack looking welds. A point that I want to stress is that you may be tempted to just pump the pedal back and forth to melt and freeze the puddle creating each individual "dime".

The problem in doing this is a couple of things. One, is you are slowing down your weld time. Meaning you're putting more heat into your work than you should be. Even though you are pulsing which lowers the actual overall heat input, you are keeping the heat on the part longer overall than you would be if you just lit up, and ran the puddle, so the part absorbs more of the heat.

Second is you are reducing your penetration. Your puddle never gets hot enough to penetrate as deeply as it should, so you end up with wider flatter beads with less penetration. you are basically just laying down a series of tack welds that never really got hot enough to penetrate deep into the base material.

Don't become a pedal pumper, you should work on arc and puddle control and learn to be able to lay down those stacks of dimes without having to vary your amperage at all. This is going to give you smooth even consistent penetration. If you are running the material at the right temperature, the puddle will flow just slow enough to hold good shape, and your filler will add just enough material to create that dime stack look without having to freeze the weld by letting off the throttle.

There are most definitely circumstances where you will need to pump the pedal to add filler to fill a gap or something, and that's fine, But I like to discourage people from relying on this technique for overall welding.

Reason being is that I especially see it on aluminum, and aluminum is notorious for weld failure, and usually that's because people aren't penetrating properly with their welds due to not enough heat. so, unless you are doing something thin and not load bearing, try to learn to work the puddle at a consistent speed and temperature.

It's all in experience, and it's more of a feel than anything. You will learn to do every weld this way, and you'll even be able to run a nice weld with the machine set a little too hot by compensating with your travel speed.

You need to be careful however with travelling too fast. There is most certainly a balance. I have seen instances where a welder is moving too fast and traps air pockets in the weld puddle, and what you end up with is a pretty looking weld with random holes in it. Nobody is going to pass a weld that looks like swiss cheese, so be careful about moving TOO fast. If you're finding this happens at a rate of speed you're comfortable at, you probably need to increase your amperage a bit

And with that I hope that you found this book useful and hopefully you picked up a couple of pointers that help you do the best job you can. Welding is one of the few trades that can be fun, aside from being rewarding. It's a great feeling when you stand back and look at your work and see how it all came together, and it's an even better feeling when someone else takes the time to notice your work and tell you what a nice job you did.

So, I decided maybe it would be helpful to someone just getting into welding, whether it be a hobbyist or an aspiring pipe liner. I wish you the best of luck on your welding adventures. It's an excellent trade to get into, it's cool to get to work on expensive projects that you'd never be able to touch on your own, but even just as a hobby, making artwork out in the garage, or building a racecar, it's still a great skill to have, and it's a great way to spend those rainy weekend nights out in the garage. So, good luck and weld on!

Reference

Filler Rod	Application
ER70s2	general steel, chromoly, will work on stainless but leaves a rusty weld.
ER70s6	General steel, chromoly, will work on stainless but leaves a rusty weld.
308	general stainless to stainless for austenitic stainless steels
316	for 316 and 316L stainless. Used in the food, medical, and restaurant industries.
312	steels to stainless. grinds easily, for welding tool steel, chromoly, and general stainless where strength is more desirable than corrosion resistance
309	Steels to stainless. carbon to 304, also works well on dissimilar steels
ERNiCr-3	Inconel - welds just about anything. Inconel and nickel alloy, Stainless to stainless, Carbon to stainless, carbon to carbon, very versatile rod, expensive.
Silicon Bronze	for brass, and TIG brazing. can be used on dissimilar materials. not for high stress joints.
4043	General aluminum fabrication works well with raw stock and most castings. lower ductility than 5353, suitable for higher temperature applications
5356	General aluminum but is more ductile than 4043. Better choice for color matching after post weld anodization. also, not suitable for higher temperature applications.

Material Quick Reference

Material	Melting Point (C)	Common Filler(s)	Process	Tensile Strength(psi)
Stainless Steel	1450	308, 316	DCEN	73200
Aluminum	660	4043, 5356	AC	45000
Carbon Steel	1500	ER70s2, s6	DCEN	53700
Chromoly	1432	Er70s2, 309, 312	DCEN	97200
Inconel 718	1425	ERNiCr-3	DCEN	128000
Brass	940	Silicon Bronze	DCEN	50000
Magnesium	650	AZ91C, AZ92A	AC	37700

Weld Joint Configuration

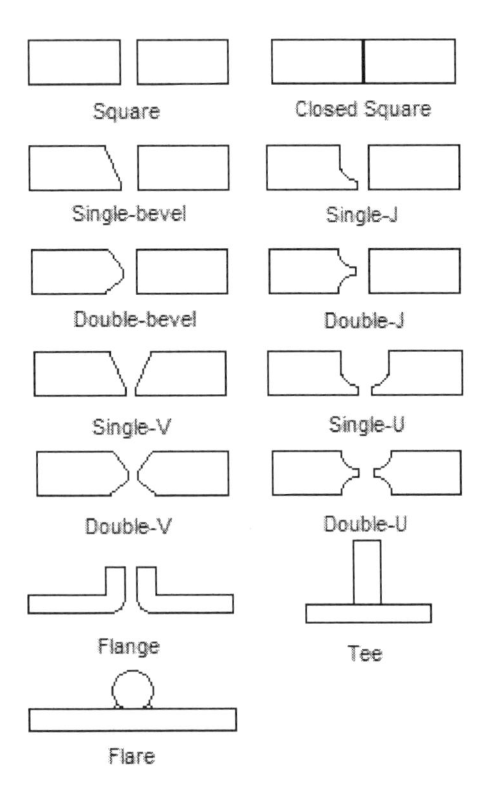

Made in United States
Orlando, FL
23 May 2025

61525840R00088